Maria Chekhova, Peter Banzer
Polarization of Light

Also of Interest

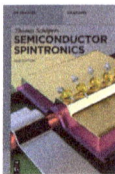

Semiconductor Spintronics
Schäpers, 2021
ISBN 978-3-11-063887-5, e-ISBN 978-3-11-063900-1

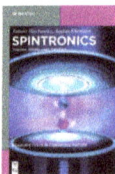

Spintronics
Theory, Modelling, Devices
Blachowicz, Ehrmann, 2019
ISBN 978-3-11-049062-6, e-ISBN 978-3-11-049063-3

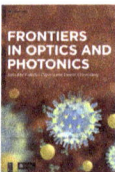

Frontiers in Optics and Photonics
Capasso, Couwenberg (Eds.), 2021
ISBN 978-3-11-070973-5, e-ISBN 978-3-11-071068-7

Quantum Electrodynamics of Photosynthesis
Braun, 2020
ISBN 978-3-11-062692-6, e-ISBN 978-3-11-062700-8

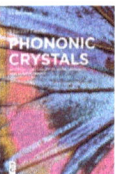

Phononic Crystals
Artificial Crystals for Sonic, Acoustic, and Elastic Waves
Laude, 2020
ISBN 978-3-11-063728-1, e-ISBN 978-3-11-064118-9

Maria Chekhova, Peter Banzer

Polarization of Light

In Classical, Quantum, and Nonlinear Optics

DE GRUYTER

Authors

Prof. Dr. Maria Chekhova
Max-Planck Institute
for the Science of Light
Staudtstr. 2
91058 Erlangen
Germany
maria.chekhova@mpl.mpg.de

Univ.-Prof. Dr. Peter Banzer
University of Graz
Universitätsplatz 5
8010 Graz
Austria

Max-Planck Institute
for the Science of Light
Staudtstr. 2
91058 Erlangen
Germany
peter.banzer@uni-graz.at

ISBN 978-3-11-066801-8
e-ISBN (PDF) 978-3-11-066802-5
e-ISBN (EPUB) 978-3-11-060509-9

Library of Congress Control Number: 2020950437

Bibliographic information published by the Deutsche Nationalbibliothek
The Deutsche Nationalbibliothek lists this publication in the Deutsche Nationalbibliografie;
detailed bibliographic data are available on the Internet at http://dnb.dnb.de.

© 2021 Walter de Gruyter GmbH, Berlin/Boston
Cover image: Andrei Berezovskii / iStock / Getty Images Plus
Typesetting: VTeX UAB, Lithuania
Printing and binding: CPI books GmbH, Leck

www.degruyter.com

For Vladimir, Rosa, Emma, Leo and Eva

Contents

1   Introduction —— 1
1.1   About this book —— 1
1.2   Brief history of polarization optics —— 2

2   Necessary basics —— 7
2.1   Analytic signal —— 7
2.2   Maxwell's equations —— 8

3   Polarization of light: classical description —— 10
3.1   Polarization ellipse —— 10
3.2   The Jones vector and the Jones matrices —— 11
3.2.1   The Jones vector —— 11
3.2.2   Different polarization states and their orthogonality —— 13
3.2.3   Different polarization bases —— 14
3.2.4   The Jones matrices —— 15
3.3   The Stokes vector and the Poincaré sphere —— 16
3.3.1   The Stokes vector —— 16
3.3.2   The Poincaré sphere —— 18
3.3.3   The Mueller matrices —— 20
3.3.4   Measurement of the Stokes observables —— 22

4   Optics of crystals: basic concepts —— 24
4.1   Anisotropy of linear optical properties —— 24
4.1.1   Dielectric tensor —— 24
4.1.2   Phase and group (ray) velocities —— 26
4.1.3   Fresnel's equation and birefringence —— 27
4.1.4   Ellipsoid of wave normals —— 29
4.2   Optical types of crystals —— 30
4.2.1   The normal surface and the ray surface —— 30
4.2.2   Crystal symmetry and the Fresnel surface —— 31
4.2.3   Circular birefringence and optical activity —— 36
4.2.4   Liquid crystals —— 38
4.3   Walk-off effects —— 38

5   Polarization transformations —— 42
5.1   Phase (retardation) plates —— 42
5.1.1   Half-wave plate —— 44
5.1.2   Quarter-wave plate —— 45
5.2   Rotators —— 45
5.3   Poincaré-sphere representation —— 47

6   Geometric phase —— 53
6.1   Examples of geometric phase —— 53
6.1.1   The Foucault pendulum —— 53
6.1.2   Non-planar optical path —— 55
6.2   Interference of arbitrarily polarized beams —— 56
6.3   Decomposition of a beam in two differently polarized
           components —— 57
6.3.1   Decomposition of a beam in two orthogonally polarized
           components —— 57
6.3.2   Decomposition of a beam in two non-orthogonally polarized
           components —— 58
6.4   Pancharatnam phase —— 59
6.4.1   Calculation of the Pancharatnam phase —— 59
6.4.2   Measurement of the Pancharatnam phase —— 60
6.5   Berry phase —— 62

7   Structured light —— 65
7.1   The paraxial wave equation —— 65
7.2   Structured scalar light beams—transverse phase patterns and phase
           singularities —— 66
7.3   Vectorial spatial modes and light beams—non-homogeneous
           polarization distributions —— 70
7.4   Polarization singularities and generic ellipse fields —— 73
7.5   Basic principles of structured light beam generation —— 75

8   Polarization of light at the nanoscale —— 79
8.1   The ubiquity of longitudinal field components —— 79
8.2   3D-Structured landscapes of light resulting from strong spatial
           confinement —— 81
8.2.1   Tight focusing —— 81
8.2.2   Near fields and evanescent waves —— 86
8.3   Measuring structured light at the nanoscale —— 87
8.3.1   Probing spatially confined fields —— 88
8.3.2   Extended Stokes parameters for three-dimensional fields —— 91
8.4   Exotic phenomena based on polarization effects in spatially confined
           light —— 93
8.4.1   Transverse spin —— 93
8.4.2   Topological features of confined light —— 97

9   Polarization elements that we use in the lab —— 101
9.1   Waveplates —— 101
9.1.1   Multiple-order plates —— 101

9.1.2   Zero-order plates —— 102
9.1.3   Dual-wavelength plates —— 103
9.1.4   Achromatic waveplates —— 103
9.1.5   Variable waveplates —— 104
9.2   Rotators —— 104
9.3   Beam displacers —— 105
9.4   Prisms and polarizing beamsplitters —— 107
9.4.1   Total internal reflection and Brewster's law —— 107
9.4.2   Glan–Taylor prism —— 108
9.4.3   Glan–Thompson prism —— 109
9.4.4   Wollaston prism —— 110
9.4.5   Polarizing cubes and plates —— 110
9.5   Fiber polarization components —— 111
9.5.1   Polarization maintaining fiber —— 111
9.5.2   Fiber polarization controllers —— 112
9.6   Liquid-crystal devices —— 112
9.6.1   Spatial light modulators —— 112
9.6.2   Q-plates —— 113

10   **Polarization in nonlinear optics —— 116**
10.1   Nonlinear susceptibilities: tensor description —— 116
10.1.1   Second-order susceptibility —— 117
10.1.2   Contracted notation —— 120
10.1.3   Various crystal symmetries —— 121
10.1.4   Third-order susceptibility —— 123
10.2   Phase matching —— 125
10.2.1   Helmholtz equation —— 125
10.2.2   Types of phase matching —— 127
10.2.3   Effective susceptibility —— 130
10.3   The effect of spatial walk-off and its elimination —— 132
10.3.1   Walk-off compensation —— 132
10.3.2   Non-critical phase matching —— 133
10.3.3   Quasi-phasematching —— 133

11   **Quantum description of polarization —— 135**
11.1   Basic notions of quantum optics —— 135
11.1.1   Modes of electromagnetic field and field quantization —— 135
11.1.2   Operators and their eigenstates —— 137
11.1.3   State characterization —— 141
11.2   Stokes observables —— 146
11.2.1   Commutation and uncertainty relations —— 147
11.2.2   A single-photon state —— 148

11.2.3   Quantum measurement of the Stokes observables —— 149
11.2.4   Weak measurement of the Stokes observables —— 153
11.3   Polarization quasi-probability —— 156

12   Nonclassical states of polarized light —— 163
12.1   Spontaneous parametric down-conversion —— 163
12.1.1   The Hamiltonian of SPDC —— 163
12.1.2   Types of phase matching for SPDC —— 167
12.2   Spontaneous four-wave mixing and related effects —— 170
12.2.1   The Hamiltonian —— 171
12.2.2   Phase matching —— 172
12.2.3   Engineering more complicated interactions —— 175
12.3   Low parametric gain and entangled photons —— 176
12.3.1   Perturbation theory: photon pairs —— 177
12.3.2   Orthogonally polarized photon pairs and polarization Hong–Ou–Mandel
            effect —— 178
12.3.3   Polarization-entangled photons and Bell states —— 182
12.4   High parametric gain: polarization squeezing and entanglement —— 183
12.4.1   Evolution of operators —— 184
12.4.2   Polarization squeezing —— 186
12.4.3   Polarization entanglement —— 189
12.5   'Hidden polarization' —— 193
12.5.1   'Hidden polarization' of nonclassical light —— 193
12.5.2   Classical 'hidden polarization' —— 195
12.5.3   Alternative definitions of the degree of polarization —— 196

13   Applications of quantum polarization states —— 200
13.1   Bell's inequality and its violation —— 200
13.1.1   The EPR paradox and Bell's inequality —— 200
13.1.2   Bell's inequality for Stokes observables —— 203
13.1.3   Bell's inequality tests —— 205
13.2   Quantum key distribution —— 208
13.2.1   Secret key —— 209
13.2.2   BB84 protocol —— 210
13.2.3   EPR-based protocols —— 212
13.2.4   B92 protocol —— 213

Index —— 217

1 Introduction

1.1 About this book

It is difficult to overestimate the role of polarization in modern optics and photonics. A brief glance at the website of any company producing optical components shows how large the section 'polarization optics' is. Polarization elements are used in imaging and spectroscopy, they can be essential in interferometers and light modulators, they are ubiquitous in lasers and laser systems. In nonlinear optics, polarization of light is crucial for understanding the phase matching and for the analysis of the tensor properties of different nonlinear susceptibilities. Polarization is important for liquid crystals, which are part of our everyday life: they are used in liquid crystal displays (LCDs) in computers and smartphones, and in spatial light modulators, installed in beam projectors. The fact that LCDs use polarization can be verified by simply looking at your mobile phone through a polarizer, or wearing polarizing sunglasses (another object familiar to everyone). As you rotate the polarizer, at a certain angle your smartphone screen will become dark.

Importantly, polarization plays the central role in modern quantum optics, quantum information, and in the booming quantum technology. The main reason for that is that the 'building bricks' of quantum information, so-called qubits, are so easily realized in the form of polarized photons. The quantum state of a polarized photon is similar to the one of a spin-1/2 particle, or of a two-level atom. Meanwhile, photons are the best carriers of information: they do not easily interact with each other or the environment; this means they can propagate relatively far without being lost or scattered. This is why polarized photons are used in quantum key distribution, one of the most robust quantum information technologies to date.

This book considers polarization of light and its manifestations and use in modern optics and photonics. It is mainly addressed to master and PhD students working in various fields of modern optics, and it is essentially based on the courses we teach at the Friedrich-Alexander University of Erlangen-Nürnberg. A large part of this book originates from the lecture course started at the Lomonosov Moscow State University by David Klyshko (and further continued by Maria Chekhova). In the quantum optics part, this book is considerably based on his work.

After introducing some necessary basics in Chapter 2, we start from the formal description of polarization (Chapter 3) in terms of the polarization ellipse, Jones vector and matrices, and Stokes vector and Müller matrices. Optics of crystals, necessary for understanding the operation of polarization optical elements, is briefly reviewed in Chapter 4. Polarization transformations with waveplates and polarization rotators are then considered in Chapter 5. Chapter 6 is devoted to the manifestations of geometric (Pancharatnam) phase in optics, similar to the Berry phase in quantum physics. Chapter 7 considers structured light, whose polarization state differs from point to point. Chapter 8 deals with polarization at the nanoscale, a subject that recently emerged in

https://doi.org/10.1515/9783110668025-001

connection with the rapidly developing fields of nanooptics and nanoscale nonlinear optics. An overview of polarization elements used in modern optics experiments is given in Chapter 9. Chapter 10 is devoted to polarization in nonlinear optics, its role in phase matching, and its manifestations due to the tensor properties of nonlinear susceptibilities. Finally, the last three chapters cover polarization-based quantum optics. Chapter 11 introduces polarization from the viewpoint of quantum physics, in terms of the Stokes operators and simplest polarization states. Chapter 12 deals with various quantum states of polarized light and related effects. Chapter 13 describes two applications of polarized light in quantum optics: one is testing the foundations of quantum mechanics, the other is quantum key distribution.

Most of the chapters are written in a textbook style and do not require special knowledge. But some of them, namely Chapters 8, 12, and 13, are also intended to give brief reviews of modern literature on the subject. Correspondingly, each of them has an extensive list of references for the interested reader. However, to keep these lists short, wherever possible we cite review papers rather than original works.

1.2 Brief history of polarization optics

Polarization of light was probably known already to ancient Vikings. There is evidence that they used $CaCO_3$ (calcite) crystal, also known as Iceland spar, for navigation. With this 'sunstone', as they called it, they managed to find the position of the sun in the sky on a cloudy day. Indeed, light is not polarized when it comes directly from the sun, but it is partially polarized when scattered [4]. This effect can be easily observed by looking at different parts of the sky through polarizing sunglasses or a polarizer: depending on the orientation of the polarizer, the sky appears brighter or darker far from the sun while it looks about the same in the area around it. But instead of a polarizer, one can use a calcite crystal due to an effect called *double refraction*. An object seen through such a crystal looks doubled; moreover, when light illuminating the object is partially polarized, the two images of the object are, in general, of different brightness. In unpolarized light, the two images will have the same brightness, no matter how one rotates the crystal. As such an object, the vikings could probably use a mark or a scratch on the external side of the calcite crystal. The direction towards the sun could be distinguished as the one in which two images of this mark were equally bright, regardless of the orientation.

Systematic study of polarization started only in the 17th century. In 1669, Bartholin observed the double refraction in calcite crystal and described it in a printed work published in 1670 [7]. Later, in 1690, Huygens declared polarization as a property of light and demonstrated it by placing two similar blocks of calcite one after another. Each crystal split an incident ray of light into two, which Huygens called 'regular' and 'irregular' [5] (ordinary and extraordinary in modern language), but if the two crystals had the same orientation, no further splitting appeared. Now we know that these ordinary and extraordinary rays are polarized orthogonally to each other, one in the plane of the crystal optic axis and the other one, perpendicularly to it. These effects of double refraction and *spatial walk-off* will be considered in detail in Chapter 4. An example of double refraction can be seen in Fig. 1.1, which shows this text on a computer screen, photographed through a 3 cm calcite crystal. The existence of only two possible polarization states,[1] for instance, vertical and horizontal, follows from light being a transverse wave; this idea was first formulated by Hooke in 1757 and further proven by Young in 1817 [7].

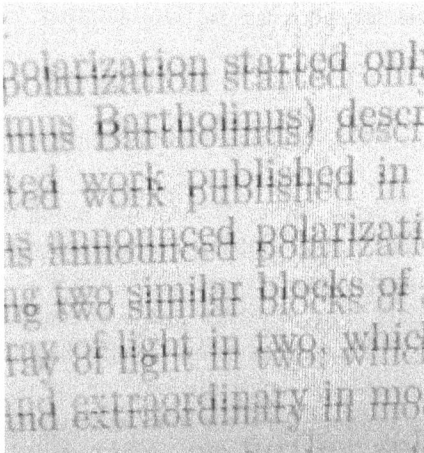

Figure 1.1: Text of this section seen on a computer screen through a 3 cm calcite crystal.

The 19th century brought enormous progress in the study of polarization. In 1808, Malus discovered that initially unpolarized light becomes partially polarized as a result of oblique reflection from a dielectric surface. The way he observed this effect was by looking through a calcite crystal at the reflections from the windows of the Luxembourg Palace in Paris, where he was an officer of the guard [4, 7]. As he rotated the crystal around its axis, one of the reflected images was extinguished. One can repeat this experiment by looking at an oblique reflection through a polarizer: at a certain orientation of the polarizer the reflected image gets weaker. Figure 1.2 shows a picture of a window in the Luxembourg Palace taken in 2019 through a polarizer selecting vertical (left) and horizontal (right) polarization. The right-hand photo obviously shows an 'extinguished' reflection. The same effect must have been seen by Malus. (Unfortunately today's guards do not let people come close to the fence, and the pictures were taken from a large distance.) The fact that for a certain angle of incidence (*Brewster's*

---

1 In some special cases, like the one of an evanescent wave, there are three possible polarization states; this will be briefly discussed in Chapter 8.

Figure 1.2: A picture of a window in the Luxembourg Palace in Paris taken through a polarizer selecting vertical (left) and horizontal (right) polarization.

angle) the horizontally polarized reflection disappears completely was established by Brewster in 1812.

Interference experiments carried out independently by Fresnel and Young in 1816–1817 led to a very important result: beams that are polarized orthogonally do not interfere. Indeed, imagine the famous Young double-slit experiment and let the polarization of light in front of one of the two slits be set as vertical and in front of the other one, horizontal. Intuitively, it is clear that there will be no interference pattern in the intensity distribution at the output. A bit less obviously, the same will be also true for the case of right- and left-circularly polarized light at the input of both slits. This is because horizontal and vertical polarization, as well as right- and left-circular polarization, form a binary set of orthogonal modes. It is this binary structure that makes polarization of light so useful for interferometry and quantum technologies.

In 1852, Stokes introduced four parameters for the description of polarization of light. They are widely used today in classical polarization optics (Chapter 3) and their counterparts, the Stokes operators, are ubiquitous in quantum polarization optics (Chapter 11). Their three-dimensional extension will be discussed in Chapter 8. The *Poincaré sphere*, which will be used intensely in this book, especially in Chapters 5, 6, 7, 12, 13, was introduced in 1892 by Poincaré. The same mathematical formalism was adopted by Bloch later, in the 20th century, for the description of a quantum system with two possible states, like a spin 1/2 particle or an atom. The *Jones vector* and *Jones matrices*, which will be used for the description of polarization in Chapter 3, were introduced by Jones in 1940. The development of the classical polarization optics formalism was accomplished in 1943 by Mueller who proposed Mueller matrices for the description of polarization transformations.

The 20th century gave birth to two new fields in optics: nonlinear optics, which emerged in the 1960s due to the advent of lasers, and quantum optics. Polarization of light played an important role in both fields. In the very first experiments by Franken and colleagues on second harmonic generation [3], it was stressed that the efficiency depends on the polarization of the pump and the orientation of the crystal. Later, the phase matching conditions were formulated, and at that time they were satisfied only through the choice of different polarization modes for the pump and the frequency converted radiation. This will be the subject of Chapter 10.

In quantum optics, polarized photons offered a possibility to realize some *gedanken* (thought) experiments formulated at the dawn of quantum mechanics. For instance, Schrödinger's concept of *entanglement* and the famous Einstein–Podolsky–Rosen (EPR) paradox of 1935 was considerably simplified by passing from the 'position-momentum' picture to the concept of a spin 1/2 particle, which was done by Bohm [1, 6]. The situation described in the EPR paradox could in this case be reproduced not in an abstract *gedanken* experiment, but in a real experiment of Stern–Gerlach type. For the latter, in 1964 Bell formulated a theorem, leading to an inequality that could help to test some statements of quantum mechanics in experiment. Although experiments on testing these *Bell's inequalities* indeed started with spin-1/2 particles, it was only through the use of *polarization-entangled photons* that Bell's inequalities could be tested in a relatively simple way. Polarization-entangled photons, as well as other types of nonclassical light, will be considered in detail in Chapter 12, and the exciting story of EPR paradox, Bell's inequalities, and their final experimental tests will be described in Chapter 13.

In 1984, Bennett and Brassard proposed an idea that later revolutionized cryptography and in fact was one of the main triggers of the quantum information theory [2]. In their method of secret key distribution between two users (now known as the *BB84 protocol*) they encoded information into the polarization states of single photons. The fragility of the polarization state of a single photon provided the protection of this secret information against eavesdropper's attacks. Generally, an information bit encoded into the state of a 'two-level' quantum system, like a photon with two polarization states, is now known as a quantum bit, a qubit, and forms the basis of quantum information. The quantum key distribution with polarized photons will be discussed in Chapter 13.

A very young field of optics, originating from the end of the 20th century, is nanooptics. It considers light at the subwavelength scale, where polarization, like many other phenomena, behaves different from a macroscopic scale. For instance, light is not any more a transverse wave: it can be polarized longitudinally. Today, with the miniaturization of optical devices, nanoscale polarization optics becomes very important. It will be the subject of Chapter 8, and the related subject of structured light beams will be considered in Chapter 7.

Bibliography

[1]   D. Bohm. *Quantum theory*. Prentice-Hall, 1952.
[2]   D. Bouwmeester, A. Ekert, and A. Zeilinger. *The physics of quantum information*. Springer-Verlag, 2000.
[3]   R. W. Boyd. *Nonlinear optics*. Academic Press, 2008.
[4]   D. Goldstein. *Polarized light*. GRC, 2003.
[5]   C. Huygens. *Treatise on light*. MacMillan and Co, 1912.
[6]   D. N. Klyshko. Basic quantum mechanical concepts from the operational viewpoint. *Phys. Usp.*, 41(9):885–922, 1998.
[7]   W. A. Shurcliff. *Polarized light*. Harvard University Press, 1962.

2 Necessary basics

2.1 Analytic signal

Although the electric field $E(t)$ is real-valued, it is very convenient to describe it by introducing a complex field, so that the observed field is its real part. To introduce this complex field [3], let us decompose the real field into a Fourier integral,

$$E(t) = \int_{-\infty}^{\infty} d\omega e^{-i\omega t} E(\omega), \tag{2.1}$$

where $E(\omega)$ is the field spectral amplitude, and split the integral in two parts, one including the integration over negative frequencies,

$$E^{(-)}(t) = \int_{-\infty}^{0} d\omega e^{-i\omega t} E(\omega), \tag{2.2}$$

and the other one, integration over positive frequencies,

$$E^{(+)}(t) = \int_{0}^{\infty} d\omega e^{-i\omega t} E(\omega). \tag{2.3}$$

The fields (2.2) and (2.3) are called negative-frequency and positive-frequency fields. They are complex conjugates of each other, as

$$[E^{(+)}(t)]^* = \int_{0}^{\infty} d\omega e^{i\omega t} E^*(\omega) = \int_{-\infty}^{0} d\omega e^{-i\omega t} E^*(-\omega)$$

$$= \int_{-\infty}^{0} d\omega e^{-i\omega t} E(\omega) = E^{(-)}(t), \tag{2.4}$$

where we have changed the integration variable from $-\omega$ to ω and used the fact that for the spectral amplitude of a real field, $E^*(\omega) = E(-\omega)$.

The positive-frequency field $E^{(+)}(t)$ will be further called the *analytic signal* [1]. The observed field is proportional to its real part,

$$E(t) = E^{(+)}(t) + E^{(-)}(t) = 2\,\mathfrak{Re}\{E^{(+)}(t)\}. \tag{2.5}$$

Figure 2.1 shows the analytic signal $E^{(+)}(t) = A(t)e^{i(-\omega t + \Phi(t))}$ for a monochromatic light. The arrow depicting the analytic signal is rotating with the optical frequency ω (dashed line), but it is convenient to consider a 'stroboscopic' picture by passing to a frame of reference rotating with the same frequency. The length of the arrow is

https://doi.org/10.1515/9783110668025-002

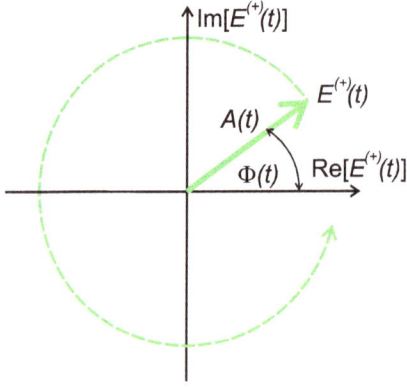

Figure 2.1: Analytic signal $E^{(+)}(t)$ (thick arrow), its amplitude $A(t)$ and phase $\Phi(t)$ in the frame rotating with the optical frequency ω.

the field amplitude $A(t)$ and its angle with the horizontal axis is the phase $\Phi(t)$. For a non-monochromatic light, both the amplitude and the phase are random functions of time. Typical times of their variation are given by the inverse spectral width of light.

The intensity of a light beam is then calculated, up to a dimensional factor, as the squared amplitude of the analytic signal. In what follows, we will omit this dimensional factor and write the instantaneous intensity as

$$I(t) = E^{(-)}(t)E^{(+)}(t) = \left|E^{(+)}(t)\right|^2. \tag{2.6}$$

2.2 Maxwell's equations

In the most general form, Maxwell's equations in SI units read [2]

$$\vec{\nabla} \times \vec{E}(\vec{r},t) = -\dot{\vec{B}}(\vec{r},t),$$
$$\vec{\nabla} \times \vec{H}(\vec{r},t) = \dot{\vec{D}}(\vec{r},t) + \vec{j}(\vec{r},t),$$
$$\vec{\nabla} \cdot \vec{D}(\vec{r},t) = \rho(\vec{r},t),$$
$$\vec{\nabla} \cdot \vec{B}(\vec{r},t) = 0. \tag{2.7}$$

Here, $\vec{r}$ is position, t is time; $\vec{E}(\vec{r},t)$ and $\vec{H}(\vec{r},t)$ are the electric and magnetic fields, respectively; $\vec{D}(\vec{r},t)$ and $\vec{B}(\vec{r},t)$ are the electric displacement and the magnetic induction, respectively; $\rho(\vec{r},t)$ is the charge density and $\vec{j}(\vec{r},t)$ the current density. The time derivative here and further throughout the book is denoted by a dot over a variable.

For the description of linear and nonlinear optical effects, Maxwell's equations should be completed by the so-called *constitutive relations*,

$$\vec{D}(\vec{r},t) = \epsilon_0 \vec{E}(\vec{r},t) + \vec{P}(\vec{r},t), \tag{2.8}$$
$$\vec{H}(\vec{r},t) = \mu_0^{-1} \vec{B}(\vec{r},t) - \vec{M}(\vec{r},t), \tag{2.9}$$

where $\vec{P}(\vec{r},t)$ is the polarization of the matter (not to be confused with the polarization of light, the subject of this book). Physically, it is the dipole moment density, i. e., the dipole moment of a unit volume, induced in the matter. Similarly, $\vec{M}(\vec{r}, t)$ is the magnetization of the matter. These values are determined by the electric and magnetic responses of the matter, namely

$$\vec{P}(\vec{r}, t) = \epsilon_0 \hat{\chi} \vec{E}(\vec{r}, t), \quad \vec{M}(\vec{r}, t) = \hat{\chi}_m \vec{H}(\vec{r}, t). \tag{2.10}$$

Here $\hat{\chi}$ and $\hat{\chi}_m$ are the electric and magnetic susceptibilities, which in the framework of this book will be assumed to be constant in space and time.

Bibliography

[1] M. Born and E. Wolf. *Principles of optics*. Pergamon Press, 1970.
[2] R. W. Boyd. *Nonlinear optics*. Academic Press, 2008.
[3] J. W. Goodman. *Statistical optics*. John Wiley and Sons, Inc., 2000.

3 Polarization of light: classical description

3.1 Polarization ellipse

Polarization of a light wave is defined by the way its electric field vector oscillates. Imagine that we can take a snapshot showing us the 'trajectory' of the electric field vector (Fig. 3.1). If the electric field $\vec{E}(\vec{r}, t)$ is oscillating in one fixed plane, light is said to be linearly polarized. If it moves along a spiral (the projection on the plane transverse to the propagation direction is a circle), the polarization state is right- or left-hand circular, depending on the direction of rotation. Finally, if the projection of the $\vec{E}(\vec{r}, t)$ trajectory on the transverse plane is an ellipse, light is elliptically polarized [6]. Figure 3.1 shows the cases of a horizontally polarized beam (a), vertically polarized beam (b), and elliptically polarized beams with different rotation direction of the electric field vector (c, d). The wave propagation direction z is the same in all pictures, and the horizontal and vertical directions are marked H and V, respectively. The pictures in the bottom row show the trajectories as seen from the direction in which the wave propagates.

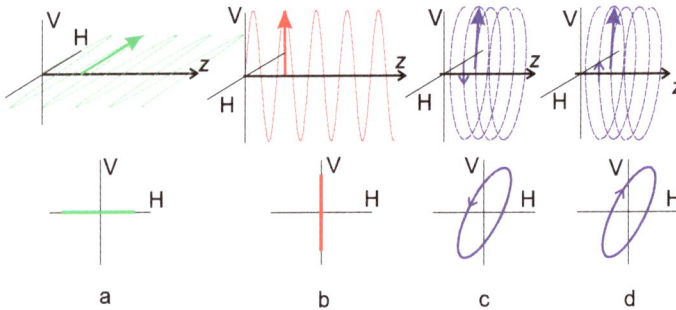

Figure 3.1: The trajectory of the electric field vector of a horizontally polarized wave (a), vertically polarized wave (b), right- (c) and left-handed (d) elliptically polarized waves. Bottom row: trajectories of the electric field vector as 'seen' by an 'observer staring into the beam'.

The latter is the most general case of a polarization state: the projection of the electric field vector trajectory on the plane orthogonal to the propagation direction can be considered as an ellipse (Fig. 3.2). The parameters of this *polarization ellipse* differ for different polarization states.

Namely, the *ellipticity* is given by the ratio of the semiaxes, b/a. Linear polarization corresponds to zero ellipticity and circular polarization, to a unity ellipticity. Instead of the ellipticity, one can speak of its opposite, the *excentricity*, $\sqrt{a^2 - b^2}/a$. The excentricity is unity for linearly polarized light and zero for circularly polarized light. The tilt angle Ω of the ellipse is called the *azimuth* angle. Finally, the direction of rotation (shown in the figure by an arrow) is called the *handedness*. It can be positive (right)

https://doi.org/10.1515/9783110668025-003

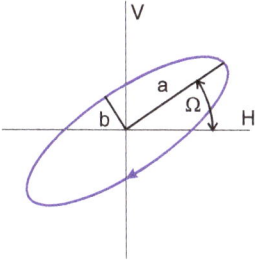

Figure 3.2: The polarization ellipse.

or negative (left). These three numbers (ellipticity, azimuth, handedness) fully define the state of polarization [6].

In Chapter 8, we will introduce another description of the polarization ellipse, which is of importance in the context of polarization singularities.

From Fig. 3.2, it is clear that the polarization state is linear when the horizontal and vertical components of the electric field oscillate in phase with each other. The azimuth is then determined by the ratio of their amplitudes. Right (left) circular polarization will be observed in the case where the horizontal and vertical components of the electric field have equal amplitudes and the latter has a phase delay of $\pi/2$ ($-\pi/2$) with respect to the former. Generally, the ellipticity is determined by the phase delay between the horizontal and vertical components of the electric field.

For monochromatic light, the phase and amplitude of the electric field (see Fig. 2.1) do not vary with time; therefore, the relative phase between the two field components, as well as the ratio of their amplitudes, is constant as well. It follows that *monochromatic light is always polarized*, i. e., its polarization state does not change in time. The situation is different for non-monochromatic light: it can happen that its polarization state drifts with time. For instance, the polarization ellipse can rotate, or its ellipticity can change. In this case, light is referred to as *partially polarized* or *unpolarized*.

Because the typical time at which the phase and amplitude of non-monochromatic light drift is the coherence time, given by the inverse width of the spectrum [3], one can conclude that at times much smaller than the coherence time even non-monochromatic light will be polarized. In particular, quite counter-intuitively, even sunlight will be polarized at very short time intervals.

3.2 The Jones vector and the Jones matrices

3.2.1 The Jones vector

The most complete classical description of the polarization state of light is in terms of the analytic signal in its vectorial form (Chapter 2). For a plane non-monochromatic wave propagating along the z axis, the analytic signal is

$$\vec{E}^{(+)}(\vec{r},t) = \vec{E}_0(t)e^{-i\omega t + ikz}, \tag{3.1}$$

where k is the wave vector, ω the central frequency, and the slowly varying amplitude $\vec{E}_0(t)$ can be decomposed in two vectors, along the horizontal (H) and vertical (V) directions:

$$\vec{E}_0(t) = \vec{E}_H(t) + \vec{E}_V(t). \tag{3.2}$$

It is convenient to define now the parameter

$$S_0(t) \equiv |E_H(t)|^2 + |E_V(t)|^2. \tag{3.3}$$

$S_0(t)$ is equal to the instantaneous intensity of the light wave. In the general case, this value varies with time [3].

The Jones vector is then defined as a two-component column vector [2, 4, 6],

$$\vec{e}(t) \equiv \begin{pmatrix} \alpha \\ \beta \end{pmatrix} \equiv \frac{1}{\sqrt{S_0(t)}} \begin{pmatrix} E_H(t) \\ E_V(t) \end{pmatrix}. \tag{3.4}$$

Here we omitted the time dependence of α and β, but we will keep in mind that both are functions of time. Therefore the Jones vector describes the *instantaneous* state of polarization.

Clearly, the Jones vector is normalized, i. e.,

$$|\alpha|^2 + |\beta|^2 = 1. \tag{3.5}$$

Since we are not interested in the overall phase of the Jones vector, but only in the relative phase between its complex components α and β, we can define them as

$$\alpha \equiv \cos(\vartheta/2),$$
$$\beta \equiv e^{i\varphi} \sin(\vartheta/2), \tag{3.6}$$

with $0 \leq \vartheta \leq \pi$ and $o \leq \varphi < 2\pi$.

We see that the polarization state is given by two numbers, the two components of the Jones vector. This has to do with the fact that the polarization state describes a *binary mode of radiation*. A radiation mode in the general case is characterized by four parameters: the three components of the wavevector, continuous variables, and polarization, a discrete variable taking two values. The concept of a polarization mode will be very important in the quantum description of light (Chapter 11).

3.2.2 Different polarization states and their orthogonality

Now we can describe different polarization states in terms of the Jones vector. For instance, for linearly polarized light, the relative phase φ between the components is zero. In particular, the Jones vector for horizontally polarized light is

$$\vec{e}_H = \begin{pmatrix} 1 \\ 0 \end{pmatrix}, \tag{3.7}$$

and for vertically polarized light,

$$\vec{e}_V = \begin{pmatrix} 0 \\ 1 \end{pmatrix}. \tag{3.8}$$

Two other important cases are the ones with $\alpha = \beta = 1/\sqrt{2}$ and with $\alpha = -\beta = 1/\sqrt{2}$. They correspond, respectively, to linear diagonal and anti-diagonal polarization states:

$$\vec{e}_{D,A} = \begin{pmatrix} 1/\sqrt{2} \\ \pm 1/\sqrt{2} \end{pmatrix}. \tag{3.9}$$

According to our considerations above, the right- and left-circularly polarized states will be the ones, for which both components of the Jones vector have equal amplitudes but phases different by $\pi/2$,

$$\vec{e}_{R,L} = \begin{pmatrix} 1/\sqrt{2} \\ \pm i/\sqrt{2} \end{pmatrix}. \tag{3.10}$$

A Jones vector with arbitrary components α, β or, equivalently, with arbitrary values of ϑ and φ, obviously describes an arbitrary elliptical polarization state. The ellipticity and azimuth of the polarization ellipse are then related to the parameters of the Jones vector by

$$\frac{a}{b} = \sqrt{\frac{1 + \sqrt{\sin^2 \vartheta \sin^2 \varphi}}{1 - \sqrt{\sin^2 \vartheta \sin^2 \varphi}}}, \quad \tan(2\Omega) = \tan \vartheta \cos \varphi. \tag{3.11}$$

The handedness is given by φ: it is right for $\varphi < \pi$ and left for $\varphi > \pi$.

One can see that the Jones vectors (3.7), (3.8), as well as two vectors (3.9) and two vectors (3.10), are orthogonal to each other. This can be verified by calculating their inner product; however, one should keep in mind that the inner product of two Jones vectors $\vec{e}_{1,2}$ with the components $\alpha_{1,2}$ and $\beta_{1,2}$ is defined as

$$\vec{e}_1 \cdot \vec{e}_2 = \begin{pmatrix} \alpha_1 \\ \beta_1 \end{pmatrix} (\alpha_2^*; \beta_2^*). \tag{3.12}$$

From this, we immediately infer that

$$\vec{e}_H \cdot \vec{e}_V = \vec{e}_D \cdot \vec{e}_A = \vec{e}_R \cdot \vec{e}_L = 0. \tag{3.13}$$

Moreover, two arbitrary Jones vectors $\vec{e}_{1,2}$ with parameters $\vartheta_{1,2}$ and $\varphi_{1,2}$ are orthogonal if $\vartheta_2 = \pi - \vartheta_1$ and $\varphi_2 = \pi + \varphi_1$. According to Eq. (3.11), this means that *for orthogonal states, the polarization ellipses have the same ellipticity, the azimuths differing by $\pi/2$ (the principal axes orthogonal), and opposite handedness.*

We see now why light beams with orthogonal polarization states (orthogonal Jones vectors) do not form an interference pattern. Indeed, if two fields overlap at some point, the total analytic signal is given by the vectorial sum of their analytic signals $\vec{E}_{1,2}^{(+)}(t)$,

$$\vec{E}^{(+)}(t) = \vec{E}_1^{(+)}(t) + \vec{E}_2^{(+)}(t), \tag{3.14}$$

and the total intensity, according to Eq. (2.6), will be given by

$$I = \left|\vec{E}_1^{(+)}(t)\right|^2 + \left|\vec{E}_2^{(+)}(t)\right|^2 + 2\,\Re\{\vec{E}_1^{(+)}(t)\vec{E}_2^{(-)}(t)\}. \tag{3.15}$$

The first two terms are the intensities of the two beams, and the last term is responsible for the interference. Its value is proportional to the inner product of the two Jones vectors, $\vec{e}_1 \cdot \vec{e}_2$. This term will be equal to zero if the Jones vectors of the two beams are orthogonal.

3.2.3 Different polarization bases

In Section 3.2.1 we decomposed the analytic signal into vertically and horizontally polarized components. This means that so far we were using the *basis* (frame of reference) formed by Jones vectors $\vec{e}_H$ and $\vec{e}_V$. We will further call it the *HV* basis. Meanwhile, in some cases it is more convenient to use other bases. We will now consider transformations from one basis to another.

Linear algebra tells us that from a basis formed by two orthonormal complex vectors $\vec{e}_{1,2}$ we can pass to a new orthonormal basis, $\vec{e}'_{1,2}$, by a transformation

$$(\vec{e}'_1, \vec{e}'_2) = (\vec{e}_1, \vec{e}_2)\mathbf{A}, \tag{3.16}$$

where $\mathbf{A}$ is a unitary matrix, $\mathbf{A}^+\mathbf{A} = \mathbf{1}$. The components of some Jones vector, initially α and β, in the new basis will become [7]

$$\begin{pmatrix} \alpha' \\ \beta' \end{pmatrix} = \mathbf{A}^+ \begin{pmatrix} \alpha \\ \beta \end{pmatrix}. \tag{3.17}$$

For example, the transformations from the HV basis to the 'diagonal' basis and 'circular' basis look, according to Eqs. (3.9) and (3.10), as

$$(\vec{e}_D, \vec{e}_A) = (\vec{e}_H, \vec{e}_V) \begin{pmatrix} \frac{1}{\sqrt{2}} & \frac{1}{\sqrt{2}} \\ \frac{1}{\sqrt{2}} & \frac{-1}{\sqrt{2}} \end{pmatrix} \tag{3.18}$$

and

$$(\vec{e}_R, \vec{e}_L) = (\vec{e}_H, \vec{e}_V) \begin{pmatrix} \frac{1}{\sqrt{2}} & \frac{1}{\sqrt{2}} \\ \frac{i}{\sqrt{2}} & \frac{-i}{\sqrt{2}} \end{pmatrix}, \tag{3.19}$$

respectively. The corresponding matrices describing the basis transformations are then

$$\mathbf{A}_{DA} = \frac{1}{\sqrt{2}} \begin{pmatrix} 1 & 1 \\ 1 & -1 \end{pmatrix}, \tag{3.20}$$

$$\mathbf{A}_{RL} = \frac{1}{\sqrt{2}} \begin{pmatrix} 1 & 1 \\ i & -i \end{pmatrix}. \tag{3.21}$$

3.2.4 The Jones matrices

Now, with the polarization state of light described by the Jones vector, it is worth discussing how this state can be transformed. We know that certain optical elements change the polarization of light. For instance, as mentioned in Chapter 1, the polarization state changes as a result of reflection. Some materials such as, for instance, the sugar solution, can rotate the plane of polarization. Another example is retardation plates used in optical laboratories: they can transform linear into circular polarization and vice versa.

In this book we will consider all these polarization elements, and some others. Our consideration will only cover lossless elements, i. e., those conserving the intensity of light. Correspondingly, these transformations will be described by unitary matrices, called the Jones matrices [2, 4]. As a result of such a transformation, an initial Jones vector $\vec{e}$ will become

$$\vec{e}\,' = \mathbf{J}\vec{e}, \tag{3.22}$$

where the *Jones matrix* $\mathbf{J}$ is a 2×2 unitary matrix. Because the total phase of the Jones vector is irrelevant[1] and is therefore ignored, the Jones matrices have an additional property $\det(\mathbf{J}) = 1$. It follows that an arbitrary Jones matrix belongs to the SU(2) group

---

1 It is, however, important for effects involving the geometric phase; see Chapter 6.

(2×2 unitary matrices with the special property of *unimodularity*, i. e., having a unity determinant) and has the general form

$$\mathbf{J} = \begin{pmatrix} t & r \\ -t^* & r^* \end{pmatrix}. \tag{3.23}$$

Here, t and r are complex numbers satisfying the condition $|t|^2 + |r|^2 = 1$.

It is worth noting that the same SU(2) form is typical for matrices describing a lossless beamsplitter: for two fields E_1 and E_2 at its input, the output fields will have the form

$$E_1' = tE_1 + rE_2,$$
$$E_2' = -t^*E_1 + r^*E_2,$$
$$|t|^2 + |r|^2 = 1. \tag{3.24}$$

Here, r and t are the reflectivity and transmissivity of the beamsplitter w. r. t. the field. This analogy between a polarization transformation and the transformation of electric fields on a beamsplitter has a simple explanation: the two polarization modes are similar to any other binary set of modes, like spatial modes, which are involved in transformations (3.24).

Thus, every polarization transforming element, like a cuvette with sugar solution, a retardation plate, or in fact any crystalline slab, will be described by a Jones matrix $\mathbf{J}$. If there are several such elements placed in series, each described by a matrix $\mathbf{J}_j$, $j = 1, \ldots, n$, then the total transformation of the Jones vector is given by the matrix $\mathbf{J} = \mathbf{J}_n \cdots \mathbf{J}_1$. In Chapter 5 we will calculate the Jones matrices of various optical elements. Polarization elements we use in the lab will be discussed in more detail in Chapter 9.

3.3 The Stokes vector and the Poincaré sphere

The Jones vector, introduced in the previous section, is directly related to the electric field and is therefore not suitable for averaging. For this reason it is not used for the consideration of partially polarized or unpolarized light. In order to describe partially polarized light, it is much more convenient to introduce polarization observables related to the intensity of light. In this section we will consider such an approach, proposed by Stokes in 1852.

3.3.1 The Stokes vector

We start by introducing the instantaneous Stokes observables, defined as

$$S_1 \equiv |E_H|^2 - |E_V|^2,$$

$$S_2 \equiv 2\,\mathfrak{Re}(E_H^* E_V),$$
$$S_3 \equiv 2\,\mathfrak{Im}(E_H^* E_V). \tag{3.25}$$

These three values form the instantaneous Stokes vector

$$\vec{S} \equiv \begin{pmatrix} S_1 \\ S_2 \\ S_3 \end{pmatrix}. \tag{3.26}$$

In addition, the zeroth Stokes observable S_0 is defined by Eq. (3.3). All Stokes observables are, in the general case, functions of time. Using the definitions (3.3), (3.25), one can verify the relation

$$S_1^2 + S_2^2 + S_3^2 = S_0^2. \tag{3.27}$$

Importantly, the second and third Stokes observables $S_{2,3}$ can be expressed similarly to S_1 by passing to the 'DA' and 'RL' bases. Indeed, because $E_{D,A} = (E_H \pm E_V)/\sqrt{2}$ and $E_{R,L} = (E_H \pm iE_V)/\sqrt{2}$, we can obtain

$$S_2 \equiv |E_D|^2 - |E_A|^2,$$
$$S_3 \equiv |E_L|^2 - |E_R|^2. \tag{3.28}$$

From this, a very simple interpretation of the Stokes observables $S_{1,2,3}$ follows: each of them is the difference of intensities in two orthogonal polarization modes,

$$S_1 \equiv I_H - I_V, \; S_2 \equiv I_D - I_A, \; S_3 \equiv I_L - I_R. \tag{3.29}$$

The mean values of the Stokes observables, $\langle S_{1,2,3}\rangle$, will be further called the *Stokes parameters*. Here we denote averaging by angular brackets, typically used in quantum mechanics, because in Chapter 11 we will apply the same notation to the quantum mechanical description. But in this chapter and further, wherever a classical description is considered, we will understand a mean value as a time-averaged quantity. Alternatively, one can imagine that the mean value is found by averaging over the ensemble formed by several independent beams.

Partially polarized light can now be described in terms of the Stokes parameters. Namely, *the degree of polarization* is introduced as

$$P \equiv \frac{\sqrt{\langle S_1\rangle^2 + \langle S_2\rangle^2 + \langle S_3\rangle^2}}{\langle S_0\rangle}. \tag{3.30}$$

For fully polarized light, the degree of polarization takes its maximum value $P = 1$ [due to Eq. (3.27)], for unpolarized light $P = 0$. In the general case, for partially polarized light, $0 \le P \le 1$.

Alternatively, unpolarized light can be described by introducing the so-called coherence matrix [1],

$$C = \begin{pmatrix} \langle E_H^* E_H \rangle & \langle E_H^* E_V \rangle \\ \langle E_H E_V^* \rangle & \langle E_V^* E_V \rangle \end{pmatrix}. \tag{3.31}$$

Then the degree of polarization (3.30) is related to the determinant and trace of the coherence matrix as [4]

$$P = \sqrt{1 - \frac{4 \det C}{(\text{Tr } C)^2}}. \tag{3.32}$$

As a result, instead of two complex variables E_H, E_V we have introduced four real variables S_0, S_1, S_2, S_3. But according to Eq. (3.27), these four numbers are not independent. It follows that by normalizing the Stokes vector $\vec{S}$ to S_0 we obtain a unit vector,

$$\vec{\sigma} \equiv \vec{S}/S_0. \tag{3.33}$$

Recalling the definition (3.4) of the Jones vector, we find that its components define the ones of the normalized Stokes vector:

$$\vec{\sigma} = \begin{pmatrix} |\alpha|^2 - |\beta|^2 \\ 2\,\mathfrak{Re}(\alpha^* \beta) \\ 2\,\mathfrak{Im}(\alpha^* \beta) \end{pmatrix}. \tag{3.34}$$

In terms of angles ϑ, φ, which we introduced in Eq. (3.6) to parametrize the Jones vector, the normalized Stokes vector takes the form

$$\vec{\sigma} = \begin{pmatrix} \cos \vartheta \\ \sin \vartheta \cos \varphi \\ \sin \vartheta \sin \varphi \end{pmatrix}. \tag{3.35}$$

3.3.2 The Poincaré sphere

Equation (3.35) is a standard notation for the spherical coordinates of a point on a unit sphere. Indeed, the variables ϑ, φ are the spherical coordinates of the normalized Stokes vector, and the corresponding unit sphere is known as *the Poincaré sphere*. Note that the orientation of the angles with respect to the axes differs from the common one: ϑ is the angle between $\vec{\sigma}$ and the σ_1 axis and φ is the angle between the projection of $\vec{\sigma}$ on the (σ_2, σ_3) plane and the σ_2 axis (Fig. 3.3).

Various points on the Poincaré sphere correspond to various polarization states. The Poincaré sphere representation provides a very convenient way to recognize these states, and it allows for using the 'geographic' terms. Consider now different parts of the sphere.

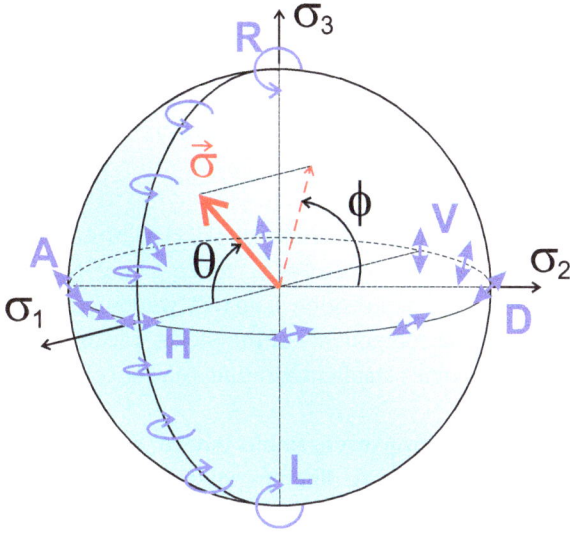

Figure 3.3: The Poincaré sphere, the normalized Stokes vector $\vec{\sigma}$ for a state with an arbitrary polarization (red arrow), and its spherical coordinates ϑ, φ. Blue arrows and ellipses show polarization states at different points on the sphere.

1. *The equator* corresponds to the case where the angle φ is either zero or π, which means that both components of the Jones vector, α and β, are real. Then there is no phase difference between the horizontal and vertical field components, i. e., the polarization is linear. Accordingly, Fig. 3.3 shows polarization states at different points of the equator as blue arrows. In particular, the point on the equator with $\vartheta = 0$ corresponds to $\alpha = 1$, $\beta = 0$, i. e., to the horizontal polarization, and the opposite point, with $\vartheta = \pi$, to $\alpha = 0$, $\beta = 1$, i. e., to the vertical polarization. Correspondingly, the points on the equator with $\vartheta = \pi/2$ and $\varphi = 0, \pi$ denote the diagonal (D) and anti-diagonal (A) polarization, respectively. As the 'longitude' on the equator increases, the plane of linear polarization tilts gradually (Fig. 3.3).

2. At the poles, $\vartheta = \pi/2$, while $\varphi = \pi/2$ for the North Pole and $\varphi = 3\pi/2$ for the South Pole. The components of the Jones vector are then $\alpha = 1/\sqrt{2}$, $\beta = i/\sqrt{2}$ for the North Pole and $\alpha = 1/\sqrt{2}$, $\beta = -i/\sqrt{2}$ for the South Pole. It follows that the North Pole depicts the right-hand circular polarization and the South Pole, the left-hand circular polarization.

   These examples, so far, show that opposite points on the Poincaré sphere correspond to orthogonal polarization states: H and V, D and A, R and L; see Fig. 3.3. This is also true in the general case: for opposite points A, B on the sphere the coordinates are $\vartheta_B = \pi - \vartheta_A$ and $\varphi_B = \pi + \varphi_A$; then, according to Eq. (3.11), opposite points on the sphere correspond to orthogonal polarization states.

3. From Eq. (3.11), it also follows that the northern hemisphere ($\varphi \leq \pi$) has right-hand polarization, while the southern one ($\pi < \varphi < 2\pi$) has left-hand polarization.

4. Points on a single meridian have the same azimuth of the polarization ellipse, equal to half of the longitude μ: $2\Omega = \mu$. In particular, for all points on the Green-

wich meridian, the azimuth is zero: the long semiaxes of the polarization ellipses are oriented along the H direction.

5. Points with the same latitude λ have the same ellipticity. From Eq. (3.11) one can see that the ellipticity is $b/a = \sqrt{(1 - \cos\lambda)/(1 + \cos\lambda)}$. In particular, for points on the equator, $\lambda = 0$, $b/a = 0$, while for the poles, $\lambda = \pi/2$ and $b/a = 1$.

Partially polarized light or unpolarized light is described in the classical picture as some variation of the Stokes vector. One can see it as the Stokes vector changing its direction with time;[2] then the point depicting the polarization state will 'wander' over the Poincaré sphere. After time averaging, the components of the Stokes vector reduce and the degree of polarization (3.30) becomes smaller than unity or, for completely unpolarized light, zero.

Alternatively, the Stokes vector can vary over an ensemble. Indeed, one can imagine that a light beam is an ensemble of beams with different polarization states. Accordingly, instead of a single point, there will be a set of points on the Poincaré sphere. To find the degree of polarization for the whole beam, one should average the Stokes vector components over all points. This procedure is equivalent to finding the vector sum of all Stokes vectors and dividing it by the number of points. As a result of this averaging, the degree of polarization becomes smaller than the unity.

3.3.3 The Mueller matrices

In section 3.2.4 we described polarization transformations in terms of the Jones vector and the Jones matrices. Because the Stokes vector, according to Eq. (3.34), is in one-to-one correspondence with the Jones vector, the same transformations can be written in terms of the Stokes vector. Indeed, as a result of a certain polarization transformation, the initial Stokes vector $\vec{\sigma}$ will become

$$\vec{\sigma}' = \mathbf{M}\vec{\sigma}, \tag{3.36}$$

where $\mathbf{M}$ is called a *Mueller matrix*. Clearly, it is a 3×3 matrix with real elements (as it relates a real vector to a real vector). From Eq. (3.34) and from the general form (3.23) of the Jones matrix we can find the corresponding Mueller matrix in terms of t and r [4]:

$$\mathbf{M} = \begin{pmatrix} |t|^2 - |r|^2 & 2\,\mathfrak{Re}(tr^*) & -2\,\mathfrak{Im}(tr^*) \\ -2\,\mathfrak{Re}(tr) & \mathfrak{Re}(t^2 - r^2) & -\,\mathfrak{Im}(t^2 + r^2) \\ -2\,\mathfrak{Im}(tr) & \mathfrak{Im}(t^2 - r^2) & \mathfrak{Re}(t^2 + r^2) \end{pmatrix}. \tag{3.37}$$

---

2 Similarly, for spatially multimode beams, the Stokes vector can be seen as varying in space.

Note that in the general case, a Mueller matrix describes a transformation of the non-normalized Stokes vector $\vec{S}$ with four components $S_{0,1,2,3}$ and it is therefore a 4×4 matrix [2, 6]. But here, we will only consider lossless transformations preserving the total intensity S_0; therefore it is sufficient to consider the normalized Stokes vector $\vec{\sigma}$ and its transformations by 3×3 matrices.

Similar to the Jones matrix, a Mueller matrix (3.37) is unitary and unimodular. It follows that a Mueller matrix conserves the length of a vector—indeed, it transforms a unit Stokes vector into another unit Stokes vector—but it also conserves the angle between two Stokes vectors. In other words, it conserves the inner product of two Stokes vectors. Moreover, one can show that it also conserves a vector product of two Stokes vectors; in particular, it transforms a 'right-hand' rectangular triplet of vectors into another 'right-hand' triplet. This means a transformation given by a Mueller matrix is a rotation on the Poincaré sphere, also known as an *SO(3) transformation* (a rotation in a three-dimensional space).

To describe such a rotation, there are at least two alternative ways. One is to specify three Euler angles. The other one, which we will follow here, is to specify the axis of rotation and the angle of rotation. They can be found from the elements of the matrix **M**. Indeed, a vector transformation (3.36) in the Cartesian space, described by a 3×3 matrix **M** with the elements m_{ij}, corresponds to the rotation of every point by an angle v about a rotation axis determined by the direction cosines c_1, c_2, c_3. (The rotation axis will be invariant to this transformation.) The rotation angle and the direction cosines are given by [5]

$$\cos v = \frac{1}{2}(\mathrm{Tr}(\mathbf{M}) - 1) = \frac{1}{2}(m_{11} + m_{22} + m_{33} - 1),$$

$$c_1 = \frac{m_{32} - m_{23}}{2 \sin v}, \quad c_2 = \frac{m_{13} - m_{31}}{2 \sin v}, \quad c_3 = \frac{m_{21} - m_{12}}{2 \sin v}. \tag{3.38}$$

From Eqs. (3.38) we obtain the angle of rotation and the direction cosines of the rotation axis for a polarization transformation in terms of t, r:

$$\cos v = \mathfrak{Re}(t^2) - |r|^2,$$

$$c_1 = \frac{-\mathfrak{Im}(t^2)}{\sin v}, \quad c_2 = \frac{2\,\mathfrak{Re}(t)\,\mathfrak{Im}(r)}{\sin v}, \quad c_3 = \frac{-2\,\mathfrak{Re}(t)\,\mathfrak{Re}(r)}{\sin v}. \tag{3.39}$$

In Chapter 5 we will consider various elements performing polarization transformations. But it is worth discussing one important case right now. Namely, consider a Jones matrix with $t = 1/\sqrt{2}$, $r = i/\sqrt{2}$. The corresponding rotation on the Poincaré sphere will be, according to Eqs. (3.39), by an angle $\pi/2$ around the σ_2 axis. Correspondingly, this transformation will convert horizontally polarized light into right-circularly polarized light, while vertically polarized light, into left-circularly polarized light. The element performing such a transformation is a quarter-wave plate (QWP, $\lambda/4$) at a certain orientation, and one needs this element for the measurement of the Stokes observables, discussed in the next section.

3.3.4 Measurement of the Stokes observables

From the definition of the Stokes observables S_0 and S_1 as the sum and difference of the intensities in the H and V polarization modes [see Eqs. (3.3) and (3.25)], it is clear how to measure these values. One should split the input beam in two, polarized horizontally and vertically, and then measure the intensities of both beams. From what was discussed in Chapter 1, this splitting can be performed with the help of a calcite crystal, but this method will be considered in detail in Chapter 4. Here we will describe a more common way, using a polarizing prism. Such a prism, also called a polarizing beamsplitter, reflects in the horizontal direction vertically polarized light and transmits horizontally polarized light. Different types of polarization prisms will be considered in Chapter 9; the operation of such a prism is shown in Fig. 3.4. After the prism, the horizontally and vertically polarized beams are measured by two detectors, and then, by summing and subtracting their readings, one can measure, respectively, the instantaneous values of S_0 and S_1.

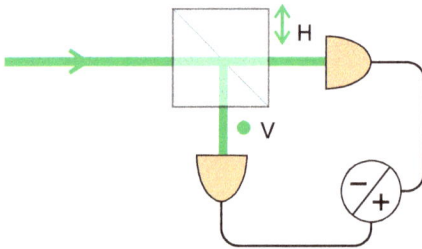

Figure 3.4: A simplified setup for the measurement of Stokes observables S_0 and S_1.

From Eqs. (3.28), one can see that the Stokes observables S_2 and S_3 can be obtained in the same way, as long as we can split the beam into two beams with D and A polarizations and into two beams with R and L polarizations. The first task is easy: one has to rotate the polarization prism by 45°. The second task is solved by using the quarter-wave plate, considered in the end of the previous section. Such a plate transforms H, V linear polarization states into R, L circular polarization states, respectively, and vice versa. Then, if it is placed in front of the polarization prism, a light beam initially polarized right-circularly will become horizontally polarized after the plate and get transmitted through the prism. Similarly, a light beam polarized left-circularly will be transformed into vertically polarized light and will be reflected. The two detectors then will measure the intensities in the R and L polarization modes (Fig. 3.5).

   With all four Stokes observables $S_{0,1,2,3}$ measured, one can calculate the normalized Stokes vector $\vec{\sigma}$. By averaging over time, the Stokes parameters and the degree of polarization can be found. In addition, it is interesting to measure the fluctuations (noise) of the Stokes observables. This is possible provided that the detectors are fast enough to follow the intensity fluctuations in each output channel of the prism. This

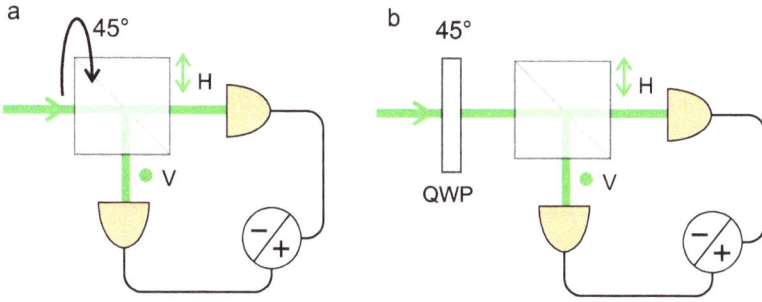

Figure 3.5: A simplified setup for the measurement of Stokes observables S_2 (a) and S_3 (b).

measurement is especially important for quantum optics and will be discussed in detail in Chapters 11 and 12.

Although we have only considered the measurement of the three Stokes observables $S_{1,2,3}$ and the total intensity S_0, it is clear that any arbitrary Stokes observable $S_{\vartheta,\varphi} \equiv S_1 \cos\vartheta + S_2 \sin\vartheta \cos\varphi + S_3 \sin\vartheta \sin\varphi$ can be also measured. The setup for this most general Stokes measurement will be considered in Chapter 5.

Bibliography

[1] M. Born and E. Wolf. *Principles of optics*. Pergamon Press, 1970.

[2] D. Goldstein. *Polarized light*. GRC, 2003.

[3] J. W. Goodman. *Statistical optics*. John Wiley and Sons, Inc., 2000.

[4] D. N. Klyshko. Polarization of light: fourth-order effects and polarization-squeezed states. *J. Exp. Theor. Phys.*, 84:1065–1079, 1997.

[5] G. A. Korn and T. M. Korn. *Mathematical handbook for scientists and engineers*. Dover Publications, Inc., 2000.

[6] W. A. Shurcliff. *Polarized light*. Harward University Press, 1962.

[7] G. Strang. *Linear algebra and its applications*. Thomson Brooks-Cole, 2006.

4 Optics of crystals: basic concepts

In this chapter, we will consider the anisotropy of optical materials, leading to bire-fringence and many other polarization effects that are discussed further in this book. We will restrict the discussion to linear optics, leaving the nonlinear effects to Chapter 10. In addition, here we will consider light at a fixed wavelength and therefore avoid the consideration of optical dispersion. We will also ignore magnetic phenomena, by assuming the matter to be non-magnetic.

4.1 Anisotropy of linear optical properties

As we will see, the linear optical properties of a material, in the general case, depend on the direction in which light propagates and, most importantly, on the polarization of light. This behavior is called anisotropy and its various aspects will be discussed in this subsection.

4.1.1 Dielectric tensor

We start with the Maxwell equations, assuming that the medium contains no charges and no currents. In this case, the general equations (2.7) simplify to (we omit the space and time arguments for brevity)

$$\vec{\nabla} \times \vec{E} = -\dot{\vec{B}}, \tag{4.1}$$

$$\vec{\nabla} \times \vec{H} = \dot{\vec{D}}, \tag{4.2}$$

$$\vec{\nabla} \cdot \vec{D} = 0, \tag{4.3}$$

$$\vec{\nabla} \cdot \vec{B} = 0. \tag{4.4}$$

In a non-magnetic material, the magnetic induction is simply proportional to the magnetic field, $\vec{B} = \mu_0 \vec{H}$. The relation between the displacement and the electric field is less trivial: Eq. (2.8) reads $\vec{D} = \epsilon_0 \vec{E} + \vec{P}$, where $\vec{P}$ is the polarization of the matter. The relation between the polarization and the electric field is provided by the first material equation of Eqs. (2.10); then

$$\vec{D} = \epsilon_0 \epsilon \vec{E}, \tag{4.5}$$

where $\epsilon = 1 + \chi$ is the *dielectric permittivity*, also known as the *dielectric function*, and χ is the linear susceptibility. We have ignored the nonlinear dependence of polarization $\vec{P}$ on the electric field $\vec{E}$. We will take it into account in Chapter 10, dedicated to nonlinear polarization optics.

https://doi.org/10.1515/9783110668025-004

Consider now the anisotropy contained in Eq. (4.5). It sets the relation between two vectors, the electric field and the displacement, through ϵ, which is therefore a tensor, ϵ_{ij}. In the matrix form, Eq. (4.5) reads

$$D_i = \epsilon_0 \sum_{j=1}^{3} \epsilon_{ij} E_j. \tag{4.6}$$

The dielectric tensor is symmetric. Indeed, the energy density of the electric field can be written as

$$U_e = \frac{1}{2} \vec{D} \cdot \vec{E} = \frac{1}{2} \epsilon_0 \sum_{i,j=1}^{3} \epsilon_{ij} E_i E_j. \tag{4.7}$$

In this expression, the indices i, j can be interchanged. This leads to the symmetry of the dielectric tensor [2]:

$$\epsilon_{ij} = \epsilon_{ji}. \tag{4.8}$$

For a generic frame of reference i, j, k, there can be six different values of the dielectric tensor: $\epsilon_{ii}, \epsilon_{jj}, \epsilon_{kk}, \epsilon_{ij}, \epsilon_{ik}, \epsilon_{jk}$. But because the energy is always positive, (4.7) is a so-called positive definite quadratic form. It is always possible to *diagonalize* it, i. e., to pass to the coordinates x, y, z such that

$$U_e = \frac{1}{2} \epsilon_0 (\epsilon_{xx} E_x^2 + \epsilon_{yy} E_y^2 + \epsilon_{zz} E_z^2). \tag{4.9}$$

The corresponding coordinate axes x, y, z are called the principal axes. In this frame of reference, the relation between the displacement and the electric field takes the simplest form

$$D_i = \epsilon_0 \epsilon_{ii} E_i. \tag{4.10}$$

The values $\epsilon_{xx}, \epsilon_{yy}, \epsilon_{zz}$ are called principal dielectric permittivities and denoted simply as $\epsilon_x, \epsilon_y, \epsilon_z$. In general, these values can change with the frequency of light and are therefore called dielectric functions. However, their symmetry should always obey the symmetry of the crystal [3] (see also Section 4.2).

It follows from Eq. (4.10) that the vectors $\vec{D}$ and $\vec{E}$ are parallel only if $\epsilon_x = \epsilon_y = \epsilon_z$. Equation (4.9) can be rewritten as

$$U_e = \frac{1}{2\epsilon_0} \left(\frac{D_x^2}{\epsilon_x} + \frac{D_y^2}{\epsilon_y} + \frac{D_z^2}{\epsilon_z} \right). \tag{4.11}$$

4.1.2 Phase and group (ray) velocities

In a plane monochromatic electromagnetic wave, all fields have harmonic space and time dependence. For the positive-frequency parts, it will be

$$E^{(+)}(\vec{r}, t) \propto e^{-i\omega t + i\vec{k}\vec{r}}, \tag{4.12}$$

and similarly for $D^{(+)}(\vec{r}, t)$, $H^{(+)}(\vec{r}, t)$, $B^{(+)}(\vec{r}, t)$. Here, ω is the frequency and $\vec{k}$ is the wavevector.

From this dependence, we can introduce the *phase velocity* $\vec{v}$. It is directed along the wavevector and its absolute value is

$$v = \frac{\omega}{k}. \tag{4.13}$$

With the refractive index n defined through the equation

$$k = \frac{n\omega}{c}, \tag{4.14}$$

the phase velocity is

$$\vec{v} = \frac{\vec{k}}{k}\frac{c}{n}. \tag{4.15}$$

With the time and space dependences of all variables taken into account, the first two Maxwell equations (4.1) and (4.2) become

$$\vec{k} \times \vec{E} = \omega\vec{B},$$
$$\vec{k} \times \vec{H} = -\omega\vec{D}. \tag{4.16}$$

At the same time, in a non-magnetic material, $\vec{B}$ and $\vec{H}$ are parallel. Then, it follows from Eqs. (4.16) that the three vectors $\vec{k}$, $\vec{E}$, $\vec{D}$ are all in the plane orthogonal to $\vec{B}$ and $\vec{H}$ (Fig. 4.1), and in this plane, $\vec{k} \perp \vec{D}$. But because $\vec{E}$ and $\vec{D}$ are not parallel, $\vec{E}$ is not orthogonal to $\vec{k}$.

At the same time, the Poynting vector, which determines the energy flux, is defined as [3]

$$\vec{S} = \vec{E} \times \vec{H}. \tag{4.17}$$

It means that the angle between $\vec{D}$ and $\vec{E}$ is the same as the angle between $\vec{k}$ and $\vec{S}$ (Fig. 4.1). This angle α is called the *angle of anisotropy* and we calculate it in Section 4.3. Similar to the vectors $\vec{E}$, $\vec{D}$, $\vec{k}$, the Poynting vector $\vec{S}$ is in the plane orthogonal to $\vec{H}$ and $\vec{B}$.

The total energy density of the electromagnetic wave is twice as large as its electric energy density (4.7), because its electric and magnetic parts are equal. Therefore,

$$U = \vec{D} \cdot \vec{E}. \tag{4.18}$$

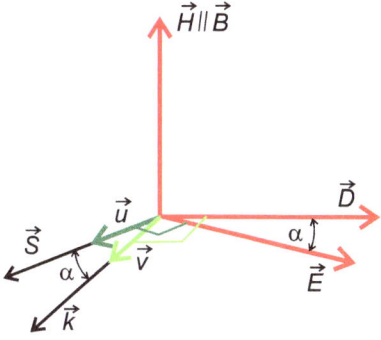

Figure 4.1: Directions of the wavevector $\vec{k}$, the Poynting vector $\vec{S}$, the field vectors $\vec{E}, \vec{D}, \vec{B}, \vec{H}$, and the phase and group velocities $\vec{v}, \vec{u}$ for a monochromatic plane wave in an anisotropic medium. Contours in the plane orthogonal to $\vec{H}\|\vec{B}$ denote right angles.

The *group velocity* of an electromagnetic wave is defined as its Poynting vector normalized to the energy density,

$$\vec{u} \equiv \frac{\vec{S}}{U} = \frac{\vec{E} \times \vec{H}}{\vec{D} \cdot \vec{E}}. \tag{4.19}$$

The group velocity is therefore directed along $\vec{S}$ and orthogonally to $\vec{E}$ by definition, and its absolute value can be found from Eq. (4.19). The absolute value of the numerator in (4.19) is EH, and of the denominator, $DE \cos \alpha$. At the same time, from the second equation of Eqs. (4.16), it follows that $D = kH/\omega$. Then,

$$\vec{u} = \frac{\vec{S}}{S} \frac{\omega}{k \cos \alpha}. \tag{4.20}$$

We see that the group velocity differs from the phase velocity only by a factor given by the cosine of the angle of anisotropy α. There is of course an additional difference between the velocities of energy and phase propagation, caused by the dispersion. However, at the moment we do not consider dispersion. For this reason, it is better to call $\vec{u}$ the *ray velocity*: it has nothing to do with the group of monochromatic waves but it shows the direction of a ray.

4.1.3 Fresnel's equation and birefringence

Let us derive the value of the phase velocity for a given direction of the wavevector. From the first equation of the system (4.16), taking into account that $\vec{B} = \mu_0 \vec{H}$, we get

$$\vec{H} = \frac{1}{\omega \mu_0} \vec{k} \times \vec{E}. \tag{4.21}$$

After substituting this expression into the second equation in (4.16) and taking into account Eq. (4.5) between $\vec{D}$ and $\vec{E}$, we obtain

$$\frac{1}{\omega^2} \vec{k} \times (\vec{k} \times \vec{E}) = -\mu_0 \epsilon_0 \epsilon \vec{E}. \tag{4.22}$$

Now, we take into account that $\mu_0 \epsilon_0 = 1/c^2$ and use the vector algebra rule

$$\vec{A} \times (\vec{B} \times \vec{C}) = \vec{B}(\vec{A} \cdot \vec{C}) - \vec{C}(\vec{A} \cdot \vec{B}). \tag{4.23}$$

We get then

$$\frac{c^2}{\omega^2}\left[\vec{k}(\vec{k} \cdot \vec{E}) - \vec{E}k^2\right] = -\epsilon\vec{E}. \tag{4.24}$$

Let us write this equation in the frame of reference where the ϵ tensor is diagonal. For each component $i = x, y, z$, we obtain

$$\frac{c^2}{\omega^2}\left[E_i k^2 - k_i \sum_{j=1}^{3} k_j E_j\right] = \epsilon_i E_i. \tag{4.25}$$

We rewrite this equation in the form

$$E_i = \frac{\sum_j k_j E_j}{k^2 - \frac{\omega^2}{c^2}\epsilon_i} k_i, \tag{4.26}$$

then multiply both parts by k_i and sum over i. We get

$$\sum_{i=1}^{3} k_i E_i = \sum_{j=1}^{3} k_j E_j \sum_{i=1}^{3} \frac{k_i^2}{k^2 - \frac{\omega^2}{c^2}\epsilon_i}. \tag{4.27}$$

The scalar product $\sum_{i=1}^{3} k_i E_i = \sum_{j=1}^{3} k_j E_j = \vec{k} \cdot \vec{E}$ on both sides is canceled, and we get

$$\frac{k_x^2}{1 - \frac{\omega^2}{c^2 k^2}\epsilon_x} + \frac{k_y^2}{1 - \frac{\omega^2}{c^2 k^2}\epsilon_y} + \frac{k_z^2}{1 - \frac{\omega^2}{c^2 k^2}\epsilon_z} = k^2. \tag{4.28}$$

In terms of the refractive index $n = ck/\omega$, the equation becomes

$$\frac{k_x^2}{1 - \frac{\epsilon_x}{n^2}} + \frac{k_y^2}{1 - \frac{\epsilon_y}{n^2}} + \frac{k_z^2}{1 - \frac{\epsilon_z}{n^2}} = k^2. \tag{4.29}$$

Let us subtract from both sides

$$k^2 = k_x^2 + k_y^2 + k_z^2. \tag{4.30}$$

We obtain

$$\frac{k_x^2}{\frac{1}{\epsilon_x} - \frac{1}{n^2}} + \frac{k_y^2}{\frac{1}{\epsilon_y} - \frac{1}{n^2}} + \frac{k_z^2}{\frac{1}{\epsilon_z} - \frac{1}{n^2}} = 0. \tag{4.31}$$

This relation is known as the Fresnel equation for wavevectors (sometimes one says 'wave normals'). After multiplying (4.31) by the product of all denominators, we obtain

$$k_x^2 \left(\frac{1}{\epsilon_y} - \frac{1}{n^2} \right) \left(\frac{1}{\epsilon_z} - \frac{1}{n^2} \right)$$
$$+ k_y^2 \left(\frac{1}{\epsilon_x} - \frac{1}{n^2} \right) \left(\frac{1}{\epsilon_z} - \frac{1}{n^2} \right)$$
$$+ k_z^2 \left(\frac{1}{\epsilon_x} - \frac{1}{n^2} \right) \left(\frac{1}{\epsilon_y} - \frac{1}{n^2} \right) = 0. \tag{4.32}$$

For any direction of $\vec{k}$, this is a quadratic equation in n^{-2}. It follows that, for any direction of $\vec{k}$, there are two values of refractive index n. This effect is called *birefringence*,[1] or *double refraction*. For instance, for $\vec{k}$ directed along the x axis, the second and third terms are zero, and the solutions are $n = \sqrt{\epsilon_y}$ and $n = \sqrt{\epsilon_z}$. In other words, for any direction of $\vec{k}$, there are two possible values of the phase velocity. For instance, for $\vec{k}$ directed along the x axis, the phase velocity can be $v = c/\sqrt{\epsilon_y}$ or $v = c/\sqrt{\epsilon_z}$.

A similar equation can be derived for the group velocity values [2].

4.1.4 Ellipsoid of wave normals

There is a clear visual explanation of the double refraction. Indeed, for a fixed electric energy of the wave, Eq. (4.11) defines an ellipsoid,

$$\frac{D_x^2}{\epsilon_x} + \frac{D_y^2}{\epsilon_y} + \frac{D_z^2}{\epsilon_z} = \text{const.} \tag{4.33}$$

By passing to the coordinates $x \sim D_x$, $y \sim D_y$, $z \sim D_z$, we get the equation

$$\frac{x^2}{\epsilon_x} + \frac{y^2}{\epsilon_y} + \frac{z^2}{\epsilon_z} = \text{const.} \tag{4.34}$$

This surface is shown in Fig. 4.2 by red color. The ellipsoid has three planes of symmetry but it is not necessarily an ellipsoid of rotation.

Let us choose a direction of the wavevector $\vec{k}$; the displacement vector $\vec{D}$ lies in the plane orthogonal to it. This plane intersects with the ellipsoid (4.2) along an ellipse, because a cross-section of an ellipsoid is always an ellipse. This ellipse is shown in orange in Fig. 4.2. One can prove [2] that the large and small semi-axes of this ellipse are along the two possible directions of $\vec{D}$ (denoted $\vec{D}'$ and $\vec{D}''$ in the figure). These are

---

1 The difference of the two refractive indices is also called birefringence.

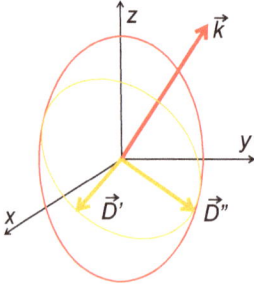

Figure 4.2: Ellipsoid of wave normals, defined by Eq. (4.34) (red), and its section by a plane orthogonal to the wavevector $\vec{k}$ (orange).

orthogonal directions. In addition, the two possible values of the phase velocity scale as the inverse lengths of these semi-axes:

$$v' \propto 1/D', \quad v'' \propto 1/D''. \tag{4.35}$$

Accordingly, the two possible refractive indices are

$$n' \propto D', \quad n'' \propto D''. \tag{4.36}$$

We arrived at an important conclusion: for the two possible values of the refractive index, the directions of displacement $\vec{D}$ are orthogonal. The three vectors $\vec{D'}$, $\vec{D''}$, and $\vec{k}$ form an orthogonal triplet of vectors.

One can also prove, in a similar way, that the two possible polarization directions, $\vec{E'}$, $\vec{E''}$, and the Poynting vector $\vec{S}$ also form an orthogonal triplet of vectors.

Optic axis or optic axes. For any ellipsoid, there are two circular cross sections passing through its center. (For an ellipsoid of rotation, these two sections coincide.) The wavevectors normal to these circular cross sections have the property that for them, $D' = D''$; hence $n' = n''$. Then there is no birefringence along these directions of $\vec{k}$, and this is the definition of an optic axis. Depending on the symmetry of the crystal, there can be one or two optic axes.

A more difficult task is to find the following surface: for each direction $\vec{k}$ one plots two wavevectors, having the two possible lengths: $k' = n'\omega/c$ and $k'' = n''\omega/c$. This surface, often called *the Fresnel surface*, has, in the general case, a more complicated shape than an ellipsoid (4.34). In the next section we consider it separately for different types of crystals.

4.2 Optical types of crystals

4.2.1 The normal surface and the ray surface

Let us find, for a given direction of the wavevector $\vec{k}$, the two possible values of the phase velocity v or the refractive index n. For this purpose, we will use Eq. (4.31). After substituting $v = \frac{c}{n}$ and denoting $v_i^2 \equiv \frac{c^2}{\epsilon_i}$, $i = x, y, z$, we get

$$\frac{k_x^2}{v^2 - v_x^2} + \frac{k_y^2}{v^2 - v_y^2} + \frac{k_z^2}{v^2 - v_z^2} = 0. \tag{4.37}$$

This dependence determines the so-called normal surface: for each direction of the wavevector two values of v are plotted. Indeed, we have shown in the previous section that there are two solutions. This surface is more complicated than an ellipse and it consists of two shells. In terms of the refractive index n, it defines the Fresnel surface.

Similarly, one can derive the equation for the ray surface. It takes the form

$$\frac{S_x^2}{u^{-2} - u_x^{-2}} + \frac{S_y^2}{u^{-2} - u_y^{-2}} + \frac{S_z^2}{u^{-2} - u_z^{-2}} = 0, \tag{4.38}$$

and it gives the two possible values of the group velocity u for a given direction of the Poynting vector $\vec{S}$.

4.2.2 Crystal symmetry and the Fresnel surface

Crystals used in linear and nonlinear optics are categorized into different classes and groups, depending on the symmetry of their elementary cell. There are seven *crystal systems*, namely, cubic, hexagonal, trigonal, tetragonal, orthorhombic, monoclinic, and triclinic, each of them containing several *crystal symmetry classes* [6, 7]. A symmetry class involves crystals whose symmetry elements (center of symmetry, mirror planes, 1-, 2-, 3-, 4-, 6-fold rotation axes or 1-, 2-, 3-, 4-, 6-fold inversion axes) belong to a certain *point group*. Cubic crystals are most symmetric, and triclinic crystals are least symmetric. There are 32 crystal symmetry classes in total, listed in Table 4.1. Here we use the so-called Hermann–Mauguin notation showing the symmetry elements of the crystals. For example, a cubic crystal of class 23 has 2-fold rotation symmetry axes parallel to the cube axes and 3-fold rotation symmetry axes parallel to the major diagonals of the cube. A cubic crystal of class 432 has 4-fold rotation symmetry axes parallel to the cube axes, 3-fold rotation symmetry axes parallel to the major diagonals of the cube, and 2-fold rotation symmetry axes parallel to the diagonals of the cube faces. For trigonal class 3m crystals, there is a 3-fold axis and three mirror planes parallel to it. A detailed explanation of this notation can be found in Ref. [6].

According to *Neumann's principle*, the symmetry of crystal physical properties in general, and optical properties in particular, should include the elements of the elementary cell symmetry. In full accordance with this statement, the optical properties of different crystal systems correspond to their symmetry; for example, cubic crystals are expected to have most symmetric optical properties.

Table 4.1: Crystal symmetry systems and their 32 classes.

System	Classes
Cubic	23; m3̄; 432; 4̄3m; m3̄m
Hexagonal	6; 6̄; 6/m; 622; 6mm; 6̄m2; 6/mmm
Tetragonal (quadratic)	4; 4̄; 4/m; 422; 4mm; 4̄2m; 4/mmm
Trigonal (rhombohedral)	3; 3̄; 32; 3m; 3̄m
Orthorhombic	222; mm2; mmm
Monoclinic	2; 2/m; m
Triclinic	1; 1̄

Depending on the symmetry, from the viewpoint of their linear optical properties, crystals can be isotropic, uniaxial, and biaxial.

The group of *isotropic crystals* only includes crystals with cubic structure. Their optical properties are the same as for isotropic materials like gases, liquids, or glasses. For such crystals, $\epsilon_x = \epsilon_y = \epsilon_z \equiv \epsilon_0$. In this case, the Fresnel equation (4.31) has only one solution, $n = \sqrt{\epsilon_0}$, and there is no birefringence.

Uniaxial crystals are less symmetric; to this class belong crystals of hexagonal, trigonal, and tetragonal systems. They have two different values of the dielectric permittivity, $\epsilon_x = \epsilon_y \equiv \epsilon_0$ and ϵ_z. In this case, the Fresnel equation (4.31), after multiplication by the product of the denominators, becomes

$$\left[\frac{1}{\epsilon_0} - \frac{1}{n^2}\right]\left[(k_x^2 + k_y^2)\left(\frac{1}{\epsilon_z} - \frac{1}{n^2}\right) + k_z^2\left(\frac{1}{\epsilon_0} - \frac{1}{n^2}\right)\right] = 0. \tag{4.39}$$

We introduce the spherical coordinates for the wavevector:

$$k_x = k\sin\vartheta\cos\varphi, \quad k_y = k\sin\vartheta\sin\varphi, \quad k_z = k\cos\vartheta. \tag{4.40}$$

After substituting this into Eq. (4.39), we find two solutions:

$$\frac{1}{\epsilon_0} - \frac{1}{n^2} = 0,$$

$$\sin^2\vartheta\left(\frac{1}{\epsilon_z} - \frac{1}{n^2}\right) + \cos^2\vartheta\left(\frac{1}{\epsilon_0} - \frac{1}{n^2}\right) = 0. \tag{4.41}$$

The solutions are

$$n = \sqrt{\epsilon_0},$$

$$n = \left(\frac{\sin^2\vartheta}{\epsilon_z} + \frac{\cos^2\vartheta}{\epsilon_0}\right)^{-1/2}. \tag{4.42}$$

These two solutions represent the Fresnel surface, shown in Fig. 4.3. The surface consists of two shells. The first one is a sphere (shown in red); for any direction of $\vec{k}$ there is the same value of the refractive index. The corresponding wave is the ordinary wave. The second one is an ellipsoid of rotation (orange in Fig. 4.3), which has a circle in the xy-cut and an ellipse in any cut passing through the z axis (because the refractive index depends only on the angle ϑ between $\vec{k}$ and the z axis). This wave is called the extraordinary wave. For a wavevector along the z axis, there is just a single value of the refractive index, $n_0 = \sqrt{\epsilon_0}$. According to the definition above, z is then the optic axis, i. e., the direction along which the two shells of the Fresnel surface intersect.

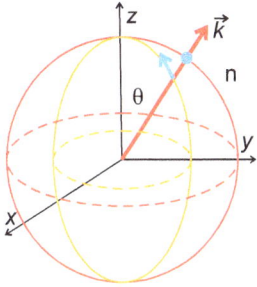

Figure 4.3: Two shells of the Fresnel surface for a uniaxial crystal, a sphere and an ellipsoid of rotation. The two possible polarization directions are shown in blue.

In other words, in a uniaxial crystal, there are two possible waves (solutions to the Maxwell equations) for any direction of the wavevector $\vec{k}$. One of them, the ordinary wave, has the refractive index independent of the $\vec{k}$ direction. For the other one, the extraordinary wave, the refractive index depends on the $\vec{k}$ direction. Along the optic axis z, both refractive indices coincide.

The directions of the electric field $\vec{E}$ for the ordinary and extraordinary waves are shown in Fig. 4.3 by a blue dot and a blue arrow, respectively. The ordinary wave is polarized orthogonally to the plane containing the wavevector and the optic axis. For the extraordinary wave, the electric field vector lies in the $(\vec{z}, \vec{k})$ plane but it is not orthogonal to $\vec{k}$. Indeed, one can show that the direction of the electric field vector is always tangent to the corresponding Fresnel surface [2]. For the ellipsoid shell of the Fresnel surface in Fig. 4.3, the tangent is not orthogonal to $\vec{k}$. This is in agreement with Fig. 4.1: in the general case, the electric field vector is not orthogonal to the wave vector.

Sometimes, n_0 is called the ordinary refractive index, n_o. The other refractive index is called the extraordinary one, $n_e \equiv \sqrt{\epsilon_z}$. A more commonly used form for the refractive index of the extraordinary wave is

$$\frac{1}{n^2} = \frac{\sin^2 \vartheta}{n_e^2} + \frac{\cos^2 \vartheta}{n_o^2}. \tag{4.43}$$

Figure 4.3 shows a situation where $n_o > n_e$. This type of uniaxial crystal is called negative; in the opposite case, $n_o < n_e$, the crystal is called positive.

The case of *biaxial crystals* is the most general one. Biaxial crystals are those belonging to triclinic, monoclinic, and orthorhombic systems. For a biaxial crystal, all values of the dielectric tensor are different, $\epsilon_x \neq \epsilon_y \neq \epsilon_z$, and the Fresnel equation has the most general form. In this case, one can say that there are two different shells of the Fresnel surface, i. e., two possible solutions to the Fresnel equation (4.31), as we have shown in Section 4.1.3. The analysis of the two Fresnel surfaces shows that they only intersect at four points, which of course should be symmetric with respect to the crystal axes. For this reason, these four points should lie in one of the principal planes, i. e., one of the planes where $k_i = 0$, $i = x, y, z$.

To find these points, consider such a plane. In this case Eq. (4.32) simplifies. For instance, for the plane orthogonal to z, we have either

$$\frac{1}{\epsilon_z} - \frac{1}{n^2} = 0, \tag{4.44}$$

or

$$k_x^2\left(\frac{1}{\epsilon_y} - \frac{1}{n^2}\right) + k_y^2\left(\frac{1}{\epsilon_x} - \frac{1}{n^2}\right) = 0. \tag{4.45}$$

The first solution defines a circle for the possible values of the refractive index n in the plane x, y: $n = \sqrt{\epsilon_z}$ regardless of the direction of $\vec{k}$. For the second solution, we get the equation of an ellipse,

$$\frac{1}{n^2} = \frac{1}{k^2}\left(\frac{k_x^2}{\epsilon_y} + \frac{k_y^2}{\epsilon_x}\right). \tag{4.46}$$

Both solutions (4.44) and (4.46), i. e., the sections of the Fresnel surfaces by the plane x, y, are shown in Fig. 4.4 by red and orange lines, respectively. The circle has the radius $n_z \equiv \sqrt{\epsilon_z}$ and the ellipse, semi-axes $n_x \equiv \sqrt{\epsilon_x}$ and $n_y \equiv \sqrt{\epsilon_y}$. The figure also shows the two polarization directions (blue arrow and blue dot), i. e., the directions of the electric field $\vec{E}$. For the circle, the electric field is orthogonal to the x, y plane and, in particular, to the $\vec{k}$ vector, as it should be for an ordinary wave. For the ellipse, light is polarized in the x, y plane (blue arrow); the electric field vector is not orthogonal

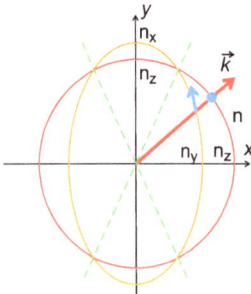

Figure 4.4: Fresnel surfaces for a biaxial crystal with $\epsilon_y < \epsilon_z < \epsilon_x$: cross-section by the xy plane. The two possible polarization directions are shown in blue.

to the $\vec{k}$ vector, similarly to the case of an extraordinary wave in a uniaxial crystal. As before, each electric field vector is tangent to the corresponding Fresnel surface.

We see that the two curves, the ellipse and the circle, intersect. Their four intersection points define the directions of the optic axes (green dashed lines in the figure). The optic axes are in the x, y plane because we assumed that $\epsilon_y < \epsilon_z < \epsilon_x$. As we will now see, the two Fresnel surfaces will not intersect in the other principal planes.

Indeed, consider now the plane orthogonal to y (shown in Fig. 4.5). Solving the Fresnel equation in this case will also lead to the equations of an ellipse and a circle for the refractive index n. The circle will now have the radius $n_y \equiv \sqrt{\epsilon_y}$, and the semi-axes of the ellipse will have larger values, $n_x \equiv \sqrt{\epsilon_x}$ and $n_z \equiv \sqrt{\epsilon_z}$. Therefore, the ellipse will be outside the circle and the two curves will not intersect. The electric field vector will be orthogonal to the xz plane in the case of the circle ('ordinary' wave, blue dot) and parallel to this plane in the case of the ellipse ('extraordinary' wave, blue arrow).

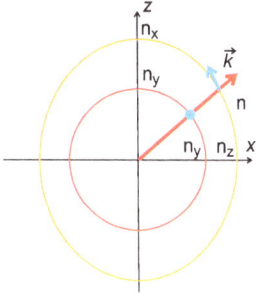

Figure 4.5: Fresnel surfaces for a biaxial crystal with $\epsilon_y < \epsilon_z < \epsilon_x$: cross-section by the xz plane. The two possible polarization directions are shown in blue.

Similarly, the third principal plane (not shown) will also intersect with the Fresnel surfaces along a circle and an ellipse; in this case the circle will have the radius n_x and lie outside of the ellipse, which will have the semi-axes n_y and n_z.

We can now combine this information about the three principal planes, each one containing an ellipse and a circle, and draw a three-dimensional plot of the Fresnel surface. A single octant of this surface is shown in Fig. 4.6. In each plane, there is a circle (red) and an ellipse (orange). As we have just seen, there is only a single intersection point within this octant belonging to the x, y plane. There are another three intersection points in the same plane, placed symmetrically with respect to x and y axes. These four points define the two optic axes, one of them shown by green dashed line in the figure.

Figure 4.6 also shows the polarization directions on each shell of the Fresnel surface. In the principal planes, the segments of the circle and the ellipse have normal and tangential polarization directions, respectively. Outside of the principal planes, polarization directions are shown on the inner Fresnel surface by red lines.

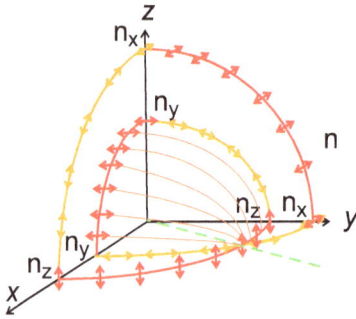

Figure 4.6: One octant of the Fresnel surfaces for a biaxial crystal with $\epsilon_y < \epsilon_z < \epsilon_x$. Polarization on each of the Fresnel surfaces is shown by arrows. The dashed green line shows the optic axis.

The orientation of the optic axes depends on the values of ϵ_x, ϵ_y, ϵ_z. In particular, if ϵ_x is close to ϵ_z, the two optic axes are close to each other and to the y axis. At $\epsilon_x = \epsilon_z$, they coincide and the crystal becomes uniaxial, with y being the optic axis.

Concluding this section, we see that in an anisotropic crystal, for each direction of the wavevector $\vec{k}$ there are two possible directions of the electric vector $\vec{E}$, both tangent to the Fresnel surface. These directions define the possible polarization states of waves that can propagate in the crystal (sometimes called *normal waves*). Normal waves are linearly polarized, and each of them has its own refractive index. The difference of these indices is called linear birefringence.

4.2.3 Circular birefringence and optical activity

After discussing linear birefringence in the previous section, we might ask a natural question: are there cases where the normal waves are elliptically or circularly polarized?

Indeed, it turns out that materials where birefringence is elliptical or circular also exist. It can take place in certain anisotropic crystals, such as uniaxial or biaxial crystals considered above. In such crystals, called *gyrotropic*, or *optically active* (this term will be clear from what follows), even in the direction of the optic axis there are still two values of the refractive index n. The two corresponding solutions to the Maxwell equations yield complex values of the analytic signal, i. e., they represent right- and left-hand circularly polarized light.

The theory of this phenomenon is based on the spatial dispersion, i. e., the dependence of the dielectric tensor on the wavevector, $\epsilon = \epsilon(\vec{k})$ [1]. Here we will not consider this theory, but only mention some examples of circular birefringence and discuss its physical consequences.

An example of a gyrotropic crystal is quartz [5]. It is a uniaxial crystal and for directions of the wavevector that are far from the optic axis it behaves as described in Section 4.2.2. The difference of the refractive indices for ordinary and extraordinary waves, $n_o - n_e$, is on the order of 10^{-2}. Meanwhile, if the $\vec{k}$ vector is directed along the optic axis, there exist, instead of a single linearly polarized wave with the refractive index n_o, two circularly polarized normal waves with refractive indices n_R (for the right-hand polarization) and n_L (left-hand polarization). The corresponding circular birefringence $n_R - n_L$ is on the order of 10^{-4}, much smaller than the linear birefringence $n_o - n_e$. In the directions close to the optic axis but not coinciding with it, normal waves are polarized elliptically. In terms of the Fresnel surfaces, optical activity means that the sphere and the ellipsoid do not touch in the direction of the optic axis, but there is a tiny gap between them [7].

Circular birefringence leads to the effect of *optical activity*, i. e., polarization rotation—and this is why gyrotropic crystals are also called optically active. A linearly polarized wave entering such a crystal will split into two circularly polarized waves, whose refractive indices, and therefore phase velocities, differ. During the propagation in the crystal, they acquire different phases and at the output, the plane of polarization is rotated. For instance, if linearly polarized light with the wavelength 500 nm is sent along the optical axis of a 3 mm quartz crystal, the plane of polarization will rotate by about 90°. This effect and the elements based on it will be considered in Chapter 5.

Circular birefringence also occurs in liquids. A well-known example is sugar solution, but many organic substances have the same property. In such substances, each molecule is *chiral*, i. e., it cannot be superposed with its mirror image. An example of a chiral object is a bolt with a thread: it is either left- or right-handed, and the two versions cannot be superposed. Similarly, a chiral molecule can be either right- or left-handed. The two different versions (*enantiomers*) do not coexist in the same solution.

The circular birefringence of organic liquids has the following explanation. For each molecule, the response to electric field is given by the *polarizability* γ: the dipole moment $\vec{d}$ induced by a field $\vec{E}$ is

$$\vec{d} = \gamma \vec{E}. \tag{4.47}$$

The polarizability of a molecule is therefore a tensor, but it is its effective value that matters for a given electric field direction. For a molecule of a certain handedness, the effective polarizability differs depending on whether the incident light is right- or left-circularly polarized: $\gamma_R \neq \gamma_L$.

The dielectric permittivity ϵ of the liquid depends on the polarizability of a single molecule according to the Lorentz–Lorenz law:

$$\frac{\epsilon - 1}{\epsilon + 2} = \frac{1}{3} N \gamma, \tag{4.48}$$

where N is the density of molecules. It follows that the dielectric permittivity also has different values $\epsilon_{R,L}$ for right- and left-circular polarization of the input light. The resulting circular birefringence $n_R - n_L$ of a liquid depends on the chiral properties of each molecule and also scales with the concentration of the molecules.

The effect of polarization rotation also depends on these parameters. Using the example of sugar, one can say that its circular birefringence is even smaller than in quartz. Correspondingly, the effect of polarization rotation is weaker: for rotating the polarization of visible light by 90°, one needs about 20 cm of saturated sugar solution.

Interestingly, in contrast to organic liquids, the optical activity of quartz is not caused by the structure of its molecules but by the crystalline structure. For instance, amorphous quartz (fused silica) does not manifest optical activity.

4.2.4 Liquid crystals

Liquid crystals are liquids that, similarly to solid crystals, manifest anisotropy. The reason for this is that their molecules are elongated.[2] For *nematic* and *smectic* liquid crystals, the molecules are axially symmetric, but in some cases (*chiral*, or *twisted*, liquid crystals) the molecules also exhibit chirality.

For such a stretched molecule, the polarizability tensor γ [see Eq. (4.47)] has at least two different components in the frame of reference of the molecule. In other words, for light polarized parallel to the long axis of the molecule (called *director*) and orthogonal to it, the response of the molecule is different.

Moreover, due to this property, the molecules can be oriented by an external static electric field. If all molecules have their directors parallel, the macroscopic optical properties are similar to the microscopic ones. Then the dielectric tensor of the whole liquid is similar to the polarizability tensor of a single molecule. For such a liquid, one can solve the Fresnel equation and find the Fresnel surfaces. For instance, in the case of a nematic or smectic liquid crystal, the Fresnel surfaces will be axially symmetric: a sphere and an ellipsoid, like for a uniaxial crystal. The optic axis direction will be given by the director of each molecule.

If light with wavevector $\vec{k}$ is incident on a cuvette with such a liquid, there will be two normal waves, ordinary and extraordinary. For the extraordinary wave, the refractive index will depend on the angle between $\vec{k}$ and the directors of the molecules. By varying the external static electric field, one can re-orient the molecules and thus change the refractive index.

These properties of liquid crystals are used in spatial light modulators and other beam-shaping devices, which will be considered in Chapters 7 and 9.

4.3 Walk-off effects

In Section 4.2.2, we saw that, for each normal wave, the direction of the electric field is tangent to the corresponding Fresnel surface. Consider, for instance, a uniaxial crys-

---

2 Sometimes one says that the molecules are cigar-shaped but this is a simplification.

tal. For the ordinary wave, the Fresnel surface is a sphere, and the electric field is orthogonal to the wavevector. But for the extraordinary wave, whose Fresnel surface is elliptic, this is not the case: the tangent to the ellipsoid is only orthogonal to the wavevector if the latter is directed along the optic axis or orthogonally to it (Fig. 4.3). For a biaxial crystal, there are also directions of the wavevector $\vec{k}$, for which the electric field for the extraordinary wave is orthogonal to it (Fig. 4.5) but this is not the general case.

Consider now the Poynting vector (4.17), which is orthogonal to the electric field and, according to Fig. 4.1, lies in the $(\vec{k}, \vec{E})$ plane. It follows that, for the ordinary wave in a uniaxial crystal, as well as for certain parts of the Fresnel surface in a biaxial crystal (see Fig. 4.6), the Poynting vector is parallel to the wavevector. This, in its turn, means that the group (ray) velocity is parallel to the phase velocity: the energy propagates in the same direction as the phase of the wave.

In contrast, for the extraordinary wave in a uniaxial crystal, unless the wavevector is parallel or orthogonal to the optic axis, the group velocity $\vec{u}$ is directed differently from the phase velocity $\vec{v}$. The angle between them, as mentioned before, is called the angle of anisotropy. This effect is called *spatial walk-off*, because the energy of the wave 'walks off' its phase propagation direction. The same walk-off effect takes place for almost all directions of the wavevector in a biaxial crystal (Fig. 4.6).

We can now explain the effect described in Chapter 1, i. e., the splitting of an image seen through a calcite crystal. If an unpolarized or arbitrary polarized beam enters a calcite crystal (Fig. 4.7) at some angle $0 < \theta < 90°$ to the optic axis, shown by a black dashed line, it splits in two beams: the ordinary one and the extraordinary one. Inside the crystal, the ordinary beam is parallel to the $\vec{k}$ vector and therefore to its direction outside. But the extraordinary beam is tilted by the anisotropy angle α inside the crystal. Outside of the crystal, there is no anisotropy and both beams are parallel again, but now they are shifted from each other by a distance $d = L \tan \alpha$, where L is the crystal length. If another crystal, with the same orientation, is placed after the first one, the shift increases. If the optic axis of the other crystal is tilted in the opposite direction, the shift is eliminated.

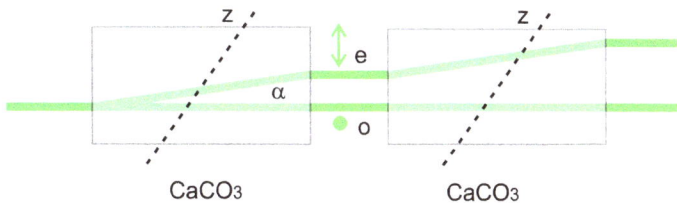

Figure 4.7: Spatial walk-off. The incoming beam has non-zero components of the polarization both perpendicular and parallel to the plane of the figure. If another crystal with a parallel optic axis is placed after the first one, the shift between the ordinary and extraordinary beams increases.

Calcite is a negative uniaxial crystal, with the refractive indices differing by about 15 %; for instance, at the wavelength 532 nm the ordinary refractive index is $n_o = 1.66$ and the extraordinary one, $n_e = 1.49$. This large birefringence leads to a very stretched Fresnel ellipsoid for the extraordinary wave (Fig. 4.3).

Taking into account that the electric field of the extraordinary wave is along the tangent to the ellipse in Fig. 4.3, we can find the anisotropy angle α from the condition

$$n \tan \alpha = \frac{\partial n}{\partial \theta}. \tag{4.49}$$

We find the derivative from Eq. (4.43):

$$\frac{\partial n}{\partial \theta} = -\frac{\sin(2\theta)n^3}{2}\left(\frac{1}{n_e^2} - \frac{1}{n_o^2}\right).$$

As a result,

$$\tan \alpha = -\frac{\sin(2\theta)n^2}{2}\left(\frac{1}{n_e^2} - \frac{1}{n_o^2}\right). \tag{4.50}$$

One can simplify it to the form [4]

$$\tan(\alpha \pm \theta) = \pm\frac{n_o^2}{n_e^2}\tan\theta, \tag{4.51}$$

where the upper signs are for negative crystals and lower signs are for positive crystals.

For the ordinary wave, $\alpha = 0$ and the Poynting vector is along the wavevector. This is why the splitting of ordinary and extraordinary beams appears. It is most pronounced, as it is clear from Eqs. (4.50), (4.51), for $\theta = 45°$. The anisotropy angle is also called the *walk-off angle*.

We have shown that spatial walk-off appears due to the non-collinearity of the phase and group velocity vectors. Similarly, *temporal walk-off* effect is caused by the difference between the absolute values of the group and phase velocities. This happens due to dispersion, i. e., to the dependence of the dielectric tensor on the frequency, which leads to the energy of an optical pulse propagating with a velocity different from that of its phase. However, in this chapter we neglect the $\epsilon(\omega)$ dependence. Still, according to Eq. (4.20), even in the absence of dispersion, the group and phase velocities differ by a factor $\cos \alpha$. This factor appears due to the non-collinearity of $\tilde{u}$ and $\tilde{v}$.

Spatial walk-off can be used for the measurement of the Stokes observables different from the one described in Section 3.3.4. This method is shown in Fig. 4.8. The light beam under study is sent to a strongly birefringent crystal (usually, calcite) with the optic axis oriented at 45°. Light polarized orthogonally to the plane formed by the wavevector and the optic axis (in the figure, horizontally) will be the ordinary beam

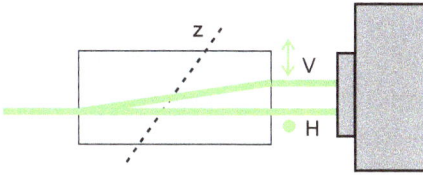

Figure 4.8: Measurement of the Stokes observables using the spatial walk-off.

and propagate straight while light polarized vertically (the extraordinary beam) will be shifted. The two output beams are registered by a camera or two detectors.

The zeroth and first Stokes observables can be measured as the sum and difference of the two intensities, respectively:

$$S_{0,1} = I_H \pm I_V. \tag{4.52}$$

With a 45° rotation, the same setup can be used to measure the second Stokes observable S_2. In order to measure the third Stokes observable S_3, one needs a quarter-wave plate in front of the setup. Its operation has been already mentioned in Chapter 3 but it will be considered in more detail in Chapter 5.

Bibliography

[1] V. M. Agranovich and V. L. Ginzburg. Crystal optics with allowance for spatial dispersion and exciton theory. *Sov. Phys. Usp.*, 5(2):323–346, feb 1962.
[2] M. Born and E. Wolf. *Principles of optics*. Pergamon Press, 1970.
[3] R. W. Boyd. *Nonlinear optics*. Academic Press, 2008.
[4] V. G. Dmitriev, G. G. Gurzadyan, and D. N. Nikogosyan. *Handbook of nonlinear crystals*. Springer, 1999.
[5] E. Hecht. *Optics*. Pearson, 2017.
[6] C. Malgrange, C. Ricolleau, and M. Schlenker. *Symmetry and physical properties of crystals*. Springer, 2014.
[7] J. F. Nye. *Physical properties of crystals*. Clarendon Press, 1985.

5 Polarization transformations

In this chapter we will consider optical elements that perform polarization transformations. We will use both the Jones vector and the Stokes vector formalisms, described in detail in Chapter 3. Here we will only consider lossless elements like waveplates and polarization rotators. Accordingly, polarizers of various types will not be the subject of this chapter but will be considered in Chapter 9.

5.1 Phase (retardation) plates

In Chapter 3 we already mentioned a quarter-wave plate. In this Section we will consider this element, as well as other types of *phase plates*, in more detail.

A *phase plate* (also called *retardation plate*) is a plate made of a lossless birefringent crystal, whose optic axis z is in the plane of the plate (Fig. 5.1) [1]. Light is incident on the plate normally to the surface (a thick red arrow in the figure). From Chapter 4 we know that, for this direction of the wavevector $\vec{k}$, only two possible normal waves can propagate inside the crystal: the ordinary and extraordinary ones. The ordinary wave is polarized orthogonally to the optic axis and has the refractive index n_o. The extraordinary wave is polarized parallel to the optic axis and has the refractive index n_e. Note that, because light is incident orthogonally to the optic axis, there is no spatial walk-off.

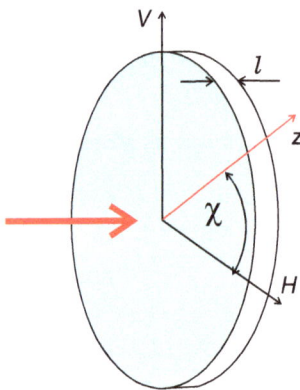

Figure 5.1: A phase plate of thickness l with the optic axis z at an angle χ to the horizontal direction.

Sometimes, when describing a waveplate, one uses the terms 'fast axis' and 'slow axis'. The *fast axis* is the polarization direction for which the normal wave has a larger phase velocity; the orthogonal direction is the *slow axis*. Usually, phase plates are made of quartz, which is a positive crystal ($n_o < n_e$), and the ordinary wave has a larger phase velocity than the extraordinary wave. Therefore the optic axis of a quartz phase plate coincides with its slow axis.

https://doi.org/10.1515/9783110668025-005

Let us calculate [1] the transformation that a phase plate of thickness l performs on the input Jones vector

$$\vec{e} = \begin{pmatrix} \alpha \\ \beta \end{pmatrix}. \tag{5.1}$$

We assume that the optic axis is at an angle χ to the horizontal axis (Fig. 5.1).

For convenience, let us first pass to the Jones vector basis associated with the plate. That is, from the original ('laboratory') basis $(\vec{e}_H, \vec{e}_V)$ we pass to the one formed by the Jones vector corresponding to the linear polarization along the optic axis, $\vec{e}_z$, and the one orthogonal to it, $\vec{e}_\perp$.

This new basis is obtained by acting on the laboratory basis with the rotation matrix $\mathbf{A}$,

$$(\vec{e}_z, \vec{e}_\perp) = (\vec{e}_H, \vec{e}_V)\mathbf{A}, \tag{5.2}$$

where

$$\mathbf{A} = \begin{pmatrix} \cos\chi & -\sin\chi \\ \sin\chi & \cos\chi \end{pmatrix}. \tag{5.3}$$

Then (see Section 3.2.3) the input Jones vector in the eigenbasis of the plate has the components

$$\begin{pmatrix} \alpha_0 \\ \beta_0 \end{pmatrix} = \mathbf{A}^+ \begin{pmatrix} \alpha \\ \beta \end{pmatrix}, \tag{5.4}$$

where we use the subscript '0' to stress that the Jones vector is written in the eigenbasis of the plate.

Inside the plate, the evolution of the Jones vector is trivial. The ordinary wave acquires a phase shift $k_o l$, where the wavevector $k_o = 2\pi n_o/\lambda$, and λ is the wavelength. Accordingly, the extraordinary wave acquires a phase shift $k_e l$, $k_e = 2\pi n_e/\lambda$. As a result, the Jones vector components after the plate become (still written in the eigenbasis of the plate)

$$\begin{pmatrix} \alpha_0' \\ \beta_0' \end{pmatrix} = \mathbf{J}_0 \begin{pmatrix} \alpha_0 \\ \beta_0 \end{pmatrix}, \tag{5.5}$$

and the Jones matrix of the plate in its eigenbasis has the form

$$\mathbf{J}_0 = \begin{pmatrix} e^{i\delta} & 0 \\ 0 & e^{-i\delta} \end{pmatrix}, \tag{5.6}$$

where $\delta \equiv \pi(n_e - n_o)l/\lambda$ and we omitted the overall phase of $\mathbf{J}_0$, $\pi(n_e + n_o)l/\lambda$, to make it an SU(2) matrix. (This phase does not change the Jones vector transformation.) As one would expect, the evolution of the Jones vector in the eigenbasis of the plate is described by a diagonal matrix.

Then, passing back to the laboratory basis, we get

$$\begin{pmatrix} \alpha' \\ \beta' \end{pmatrix} = (\mathbf{A}^+)^{-1} \begin{pmatrix} \alpha'_0 \\ \beta'_0 \end{pmatrix}. \tag{5.7}$$

It follows that the transformation performed by the phase plate is written in the laboratory frame as

$$\begin{pmatrix} \alpha' \\ \beta' \end{pmatrix} = \mathbf{J}_p \begin{pmatrix} \alpha \\ \beta \end{pmatrix}, \tag{5.8}$$

where

$$\mathbf{J}_p = (\mathbf{A}^+)^{-1} \mathbf{J}_0 \mathbf{A}^{-1} = \mathbf{A} \mathbf{J}_0 \mathbf{A}^+. \tag{5.9}$$

Here, we used the unitarity of the matrix $\mathbf{A}$.

By multiplying the matrices, we get the Jones matrix of a phase plate:

$$\mathbf{J}_p = \begin{pmatrix} \cos\delta + i\sin\delta\cos(2\chi) & i\sin\delta\sin(2\chi) \\ i\sin\delta\sin(2\chi) & \cos\delta - i\sin\delta\cos(2\chi) \end{pmatrix}. \tag{5.10}$$

This expression can be written in the general form (3.23) of an SU(2) matrix, with $t = \cos\delta + i\sin\delta\cos(2\chi)$ and $r = i\sin\delta\sin(2\chi)$.

A generic Jones matrix (3.23) is given by two complex numbers t, r, which boil down, taking into account the condition $|t|^2 + |r|^2 = 1$, to three real parameters. From the structure of the Jones matrix (5.10) for a phase plate, we see that it has only two parameters, the phase δ and the angle χ. Therefore, a general type of polarization transformation cannot be realized by a single phase plate, but only by a combination of two plates.

In the next section, we consider the most commonly used types of phase plates.

5.1.1 Half-wave plate

For a half-wave plate (HWP), the phase $\delta = \pi/2$. According to Eq. (5.6), the ordinary and extraordinary waves acquire in this plate a relative phase of π. The Jones matrix of such a plate is then

$$\mathbf{J}_{\text{HWP}} = i \begin{pmatrix} \cos(2\chi) & \sin(2\chi) \\ \sin(2\chi) & -\cos(2\chi) \end{pmatrix}. \tag{5.11}$$

Such a plate will leave a linearly polarized beam linearly polarized but it will rotate the polarization direction. Indeed, consider the Jones vector of linearly polarized light,

$$\vec{e} = \begin{pmatrix} \cos(\vartheta/2) \\ \sin(\vartheta/2) \end{pmatrix}, \tag{5.12}$$

with the initial azimuth $\Omega = \vartheta/2$. After a HWP oriented at an angle χ, the Jones vector becomes, up to a phase factor i,

$$\vec{e}' = \begin{pmatrix} \cos(2\chi - \vartheta/2) \\ \sin(2\chi - \vartheta/2) \end{pmatrix}. \tag{5.13}$$

We see that light is still polarized linearly but the azimuth is now $2\chi - \vartheta/2$. In particular, if the HWP is oriented at an angle $\chi = \pi/4$, the polarization plane will be rotated by $90°$ with respect to its initial state.

Typically, half-wave plates are used in laboratories to rotate the polarization plane. However, this operation is only possible with linearly polarized light. As we will see in Section 5.3, the way a HWP acts on elliptically or circularly polarized light is different. In particular, it transforms right-hand circularly polarized light into left-hand cicularly polarized light and vice versa.

5.1.2 Quarter-wave plate

For a quarter-wave plate (QWP), the phase $\delta = \pi/4$. The Jones matrix of such a plate is

$$\mathbf{J}_{\text{QWP}} = \frac{1}{\sqrt{2}} \begin{pmatrix} 1 + i\cos(2\chi) & i\sin(2\chi) \\ i\sin(2\chi) & 1 - i\cos(2\chi) \end{pmatrix}. \tag{5.14}$$

Although in the general case the transformation performed by a QWP is quite complicated, it is mostly used to convert linearly polarized light into circularly polarized light and vice versa. In this case, the orientation of the QWP should be $\chi = \pi/4$. Then, if light at the input of a QWP is horizontally polarized, i. e. its Jones vector has the form (3.7), after the QWP the Jones vector is

$$\vec{e}' = \frac{1}{\sqrt{2}} \begin{pmatrix} 1 \\ i \end{pmatrix}, \tag{5.15}$$

i. e., light is right-hand circularly polarized [compare with Eq. (3.10)]. With the vertically polarized light at the input of a QWP, light at its output will be left-hand circularly polarized.

5.2 Rotators

A polarization rotator is a device rotating the polarization plane of linearly polarized light [1, 4]. Such an element can be constructed using an optically active material (see Section 4.2.3), for instance a cuvette with sugar solution or a gyrotropic crystal. At first sight, a rotator performs the same operation as a HWP. However, this is not the case, as we will soon see.

As described in Section 4.2.3, the eigenstates (normal waves) in an optically active material are circularly polarized. To describe the transformation of the Jones vector, we again pass from the laboratory basis to the eigenbasis of the rotator. This will be the circular basis $(\vec{e}_R, \vec{e}_L)$, given by Eq. (3.10). The transformation from the laboratory basis to the circular one is given by the matrix $\mathbf{A}_{RL}$; see Eq. (3.21). The Jones matrix in the eigenbasis will be again diagonal, of the form (5.6); however, now, unlike in the example with the plates, the birefringence is circular: $\delta \equiv \pi(n_R - n_L)l/\lambda$.

The resulting Jones matrix, similar to the case of the phase plate [see Eq. (5.9)], is calculated as

$$\mathbf{J}_r = \mathbf{A}_{RL}\mathbf{J}_0\mathbf{A}_{RL}^+. \tag{5.16}$$

As a result, we obtain

$$\mathbf{J}_r = \begin{pmatrix} \cos\delta & \sin\delta \\ -\sin\delta & \cos\delta \end{pmatrix}. \tag{5.17}$$

This is a rotation matrix, with the angle of rotation δ. For instance, if horizontally polarized light is at the input of the rotator, at the output the Jones vector becomes

$$\vec{e}' = \begin{pmatrix} \cos\delta \\ \sin\delta \end{pmatrix}, \tag{5.18}$$

i. e., the initial polarization azimuth is rotated by an angle δ. Clearly, Eq. (5.17) is also an SU(2) matrix, characterized by a single parameter δ.

Now we can see how the operation of a rotator is different from the one of a HWP. For linearly polarized light, the former rotates the polarization by the same amount, regardless of its orientation. The latter, meanwhile, rotates the polarization depending on its orientation. In particular, if the input light is polarized along the fast or slow axis of a HWP, no polarization rotation occurs: these states are eigenstates of the HWP. For a rotator, the eigenstates are circularly polarized; therefore, it leaves circularly polarized light unchanged.

An example of a rotator is a quartz crystal cut orthogonally to the optic axis. Figure 5.2 shows the thickness of a quartz slab needed to rotate linear polarization by 45° (red line) and 90° (blue line) for different wavelengths [3]. The optical activity of quartz is especially strong in the ultraviolet (UV) range, where a 100μ-thick plate already rotates polarization by a few tens of degrees. Such a device is convenient because the rotation is performed regardless of the initial polarization, unlike in the case of a HWP. The drawback of a quartz polarization rotator is that the angle of rotation depends on the wavelength, which can be a problem when dealing with broadband light.

Apart from the optically active material like crystal quartz or a sugar solution, a rotator can be constructed using the Faraday effect, i. e., rotation of the plane of polarization in a magnetic field. This way one obtains a Faraday cell, which will be considered in more detail in Chapter 9. The most important feature of this type of rotator

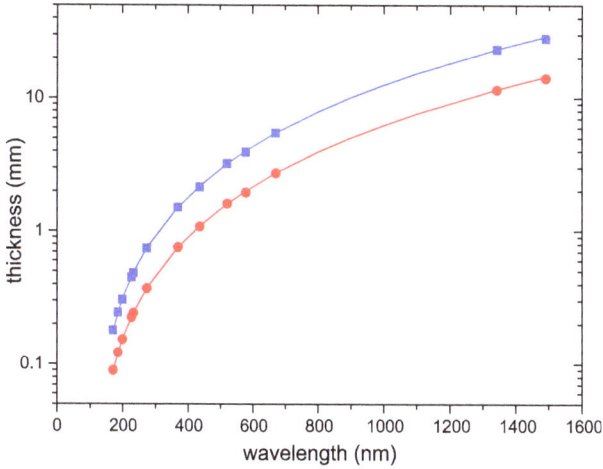

Figure 5.2: Thickness of a quartz plate cut orthogonal to the optic axis, needed to rotate the polarization by 45° (red) and 90° (blue) [3].

is that its phase δ depends on the direction of propagation of light, which is not the case for an optically active material.

Let us briefly mention here another, very elegant way of rotating the plane of polarization for linearly polarized light. This method consists of reflecting a light beam in three different planes, so that the final direction of the beam coincides with the initial one. Such triple reflection, in the general case, leads to a polarization transformation. In particular, if reflections occur in three mutually orthogonal planes, light remains linearly polarized but its plane of polarization is rotated by $\pi/2$. This method is based on the effect similar to the geometric phase and will be considered in Chapter 6, and the corresponding device, consisting of three mirrors, is called an *optical tower* or a *periscope*.

5.3 Poincaré-sphere representation

As mentioned in Section 3.3.3, polarization transformations described by Jones matrices correspond to rotations of the Stokes vector on the Poincaré sphere. In this section, we will find these rotations using Eqs. (3.39) for the angle of rotation v and the direction cosines c_1, c_2, c_3 that define the axis of rotation [2].

For a phase plate described in Section 5.1, the elements of the Jones matrix are $t = \cos\delta + i\sin\delta\cos(2\chi)$, $r = i\sin\delta\sin(2\chi)$. Then, from Eqs. (3.39), we find that the angle of rotation v is given by

$$\cos v = \cos(2\delta), \tag{5.19}$$

and the direction cosines for the axis of rotation are

$$c_1 = -\cos(2\chi), \quad c_2 = -\sin(2\chi), \quad c_3 = 0. \tag{5.20}$$

It follows that a plate with the phase δ and the orientation angle χ rotates the Stokes vector by an angle 2δ around an axis that lies in the equatorial plane at an angle 2χ to the σ_1 axis. Figure 5.3 shows this rotation on the Poincaré sphere by a red arrow; the axis of rotation is shown by a red dotted line.

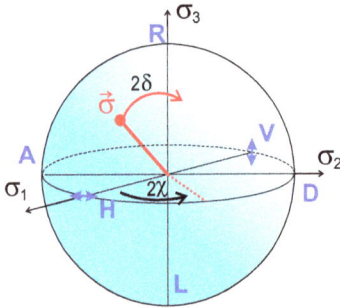

Figure 5.3: Rotation of the normalized Stokes vector (red point on the Poincaré sphere) performed by a phase plate.

Note that the same rotation can be considered in two different ways. For instance, Fig. 5.3 shows a point on the Poincaré sphere rotated clockwise by an angle 2δ. This is the so-called *active viewpoint*. But the same effect will be achieved if the point stays the same, but the sphere rotates anticlockwise by an angle 2δ. Sometimes this *passive viewpoint* is more convenient.

A *half-wave plate* will always rotate the Stokes vector by an angle π. It means, in agreement with the result we obtained in Section 5.1.1, that it will always transform linearly polarized light (a point on the equator of the Poincaré sphere) into linearly polarized light (another point on the equator). For instance, if the initial polarization state is horizontal ($\vartheta = 0$), after the rotation we get $\vartheta = 4\chi$, i.e., the final state is linearly polarized at an angle 2χ. (Note that angles on the Poincaré sphere are always doubled compared to their values in the usual space.) These rotations are shown in Fig. 5.4 by different colors: the HWP is oriented at 11.25° (orange), at 22.5° (green), and at 45° (purple). The latter is the most commonly used setting of the HWP: when it is oriented at $\pi/4$ to the incident light polarization, it rotates the polarization by $\pi/2$. The axis of rotation in each case is plotted by a dotted line of the corresponding color.

If the state is originally circularly polarized, it corresponds to one of the poles on the Poncaré sphere, and the HWP will transform it into the other pole, regardless of the orientation χ of the plate. In Fig. 5.4, the magenta line shows the transformation a HWP performs on right-hand circularly polarized light. The trajectory, however, will go along different meridians, depending on the angle χ (for the example in Fig. 5.4,

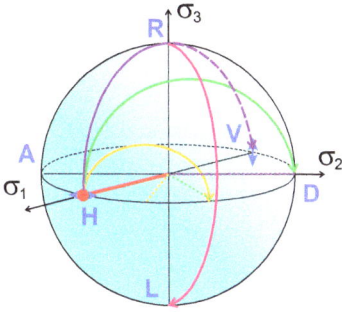

Figure 5.4: Rotation of the normalized Stokes vector of a linearly polarized state (red dot) performed by a HWP. Different orientations of the HWP correspond to different colours: 11.25° (orange), 22.5° (green), 45° (purple). The magenta line shows rotation of the Stokes vector of a right-hand circularly polarized state with a HWP oriented at about −10°.

$\chi \approx -10°$). This fact will be important in Chapter 6, where we discuss the geometric phase.

A *quarter-wave plate* will rotate the Stokes vector by an angle $\pi/2$. It is clear now why it transforms linearly polarized light into circularly polarized light. For instance, if the initial polarization state is horizontal (Fig. 5.5, red dot) and the plate is oriented at $\chi = \pi/4$, the rotation is around the σ_2 axis and the final state is right-hand circularly polarized (solid purple arrow in Fig. 5.5). Similarly, if the initial state is vertically polarized, light will be left-hand circularly polarized after the plate (dashed purple arrow in Fig. 5.5). The same transformation will turn circularly polarized light into linearly polarized light: R into V, L into H. Other transformations with the QWP are shown in the figure by orange (QWP at 11.25°) and green (QWP at 22.5°) colors.

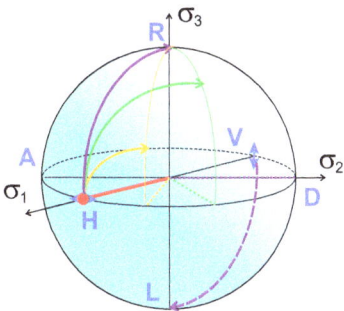

Figure 5.5: Rotations performed on the Poincaré sphere by a QWP. The initial polarization state is horizontal, and the orientations of the QWP are as in Fig. 5.4: 11.25° (orange), 22.5° (green), 45° (purple).

It is worth mentioning that, obviously, two QWPs with the same orientation χ, stacked together, form a HWP. This simply follows from the fact that the phase δ is additive, and it will be $\pi/2$ for the system of two QWPs. The same conclusion follows if we consider two consecutive rotations by $\pi/2$ around the same axis: obviously it is a π rotation around the same axis. These simple considerations explain a very convenient method used in many polarization setups: if a linearly polarized beam passes through a QWP oriented at $\pi/4$ to its polarization plane and then is reflected by a mirror at normal incidence, then its polarization, after passing the QWP twice, will be rotated by $\pi/2$.

If the QWP is oriented at an angle smaller than $\pi/4$, then the initial point H will be transformed into a point with a lower latitude (orange and green arrows in Fig. 5.5), and the resulting polarization will be elliptical. In the opposite situation, where light of given ellipticity has to be transformed into linearly polarized light, one can also use a QWP: the required orientation χ is then determined by the ellipticity.

A plate with an arbitrary phase δ will perform a rotation by an arbitrary angle, but always around an axis lying in the equatorial plane. This is a restriction imposed by the fact that, for any plate, polarization eigenstates are linearly polarized. As mentioned above, this does not allow for performing an arbitrary SU(2) transformation with a single plate. At the same time, if the goal is to transform a given initial polarization state into another polarization state, it is always possible to find a plate performing such a transformation. Note also that a transformation from one point on the Poincaré sphere into another point can be performed by infinitely many rotations, only one of them being possible with a phase plate.

A combination of several plates will result in a combination of rotations, the total Mueller matrix being the product of the matrices for all plates. The best-known example is a combination of three plates: QWP + HWP + QWP. For an arbitrary input polarization state, the first plate (QWP) is used to transform it into linear polarization, then a HWP rotates the linear polarization by a necessary amount, and the last plate (another QWP) produces a state with a given ellipticity. The same system can be realized with fiber loops (which will be discussed in more detail in Chapter 9). A simplified version, a combination of a QWP and a HWP, transforms an arbitrary state into an arbitrary linearly polarized state and is used in the measurement of an arbitrary Stokes observable (Fig. 5.7).

Transformations with a polarization rotator are considered similarly [2]. From Eqs. (3.39), we find that, for a rotator with the phase δ, the rotation is by the angle v, with

$$\cos v = \cos(2\delta), \tag{5.21}$$

and the axis of rotation is given by its direction cosines

$$c_1 = c_2 = 0, \quad c_3 = -1. \tag{5.22}$$

This rotation is around the σ_3 axis, by an angle 2δ (Fig. 5.6). A rotator can then move any point along the equator, and it would leave the poles (right- and left-circular polarization states) unchanged.

The reason why a circularly polarized state stays invariant under the action of a rotator is because this is an eigenstate of the rotator. Similarly, in the case of a waveplate, if the incident light beam is polarized linearly along the optic axis of the waveplate, the rotation on the Poincaré sphere is around the initial point itself. This rotation obviously leaves the polarization state the same. Note that, if light is initially polarized orthogonally to the optic axis of the plate, the point depicting its polarization state

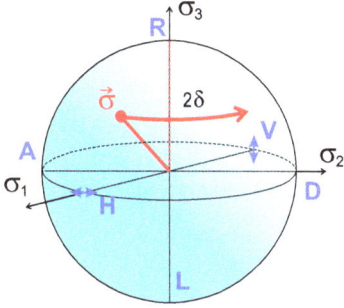

Figure 5.6: Rotation performed on the Poincaré sphere by a rotator.

on the Poincaré sphere lies on the other side of the same diameter, and therefore this polarization state is also left unchanged under this transformation.

We see that the Poincaré sphere is a very convenient tool to visualize polarization transformations. It is also helpful to understand the transitions between the measurements of different Stokes observables. Indeed, from the structure of the Poincaré sphere we see that there is no principal difference between the Stokes observables σ_1, σ_2, σ_3 and any generalized Stokes observable

$$\sigma_{\vartheta,\varphi} \equiv \cos\vartheta\,\sigma_1 + \sin\vartheta\cos\varphi\,\sigma_2 + \sin\vartheta\sin\varphi\,\sigma_3. \tag{5.23}$$

In order to switch from the measurement of the first Stokes observable to the measurement of the second Stokes observable, we need to rotate our frame of reference so that the σ_2 axis is transformed into the σ_1 axis. This rotation of the frame of reference can be performed by placing a HWP with $\chi = \pi/8$ in front of the measurement scheme in Fig. 3.4. From the active viewpoint, it rotates the input linear polarization by $\pi/4$, transforming A into H and D into V. But from the passive viewpoint [2], it transforms the σ_2 axis into the σ_1 axis and therefore enables the measurement of the second Stokes observable. Note that the same transformation is achieved by means of a rotator with $\delta = \pi/4$.

Similarly, for the measurement of the third Stokes observable we need to transform the σ_3 axis into the σ_1 axis. This transformation is performed by a QWP set at an angle $\chi = \pi/4$. This is why this plate is used in the setup shown in Fig. 3.5 for the measurement of S_3.

It follows that an arbitrary Stokes observable (5.23) can be measured with a setup where the polarization prism is preceded by a set of waveplates performing a polarization transformation of a general form. Because the final polarization basis is set by the prism, a combination of only two plates suffices, QWP + HWP. By setting their orientation angles χ_1 and χ_2, any values of ϑ, φ can be achieved. Figure 5.7 shows this general scheme of the Stokes measurement. By propagating the Stokes vectors 'backwards' through the setup, we see that the Stokes vector $\vec{\sigma}_{\text{out}}$ with $\sigma_1 = 1$, $\sigma_2 = \sigma_3 = 0$ at the output corresponds to the input Stokes vector

$$\vec{\sigma}_{\text{in}} = \mathbf{M}_1\mathbf{M}_2\vec{\sigma}_{\text{out}}, \tag{5.24}$$

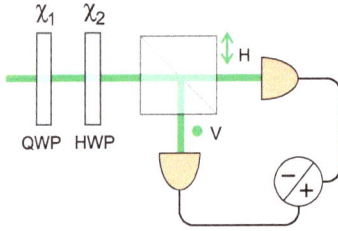

Figure 5.7: Measurement of an arbitrary Stokes observable.

where $\mathbf{M}_1$, $\mathbf{M}_2$ are the Mueller matrices of the QWP and HWP, respectively. Calculation yields for $\vec{\sigma}_{in}$ the spherical angles ϑ and φ given by

$$\cos\vartheta = \cos(2\chi_1)\cos(2\chi_1 - 4\chi_2), \quad \tan\varphi = -\frac{\tan(2\chi_1 - 4\chi_2)}{\sin(2\chi_1)}. \tag{5.25}$$

The setup shown in Fig. 5.7 can be used to measure the degree of polarization. According to the definition (3.30), it is given by the visibility of the intensity modulation obtained at one of the outputs under all possible rotations of the QWP and HWP [2]:

$$P = \frac{I_{max} - I_{min}}{I_{max} + I_{min}}, \tag{5.26}$$

where I_{max} and I_{min} are, respectively, the maximal and minimal intensity measured by one of the detectors of the setup in Fig. 5.7.

As a final comment, let us mention that apart from rotations, which we considered so far, one might also consider other transformations, like mirror reflections and inversion. Inversion, which brings every point on the sphere into the opposite point, is especially interesting for polarization optics because it would transform every polarization state into an orthogonal one. Unfortunately, inversion or mirror reflection on the Poincaré sphere is not possible using only phase plates or rotators because these elements perform only rotations. Any rotation leaves the points on the axis of rotation invariant and therefore cannot lead to their inversion or mirror reflection.

Bibliography

[1] D. N. Klyshko. Berry geometric phase in oscillatory processes. *Phys. Usp.*, 36(11):1005–1019, nov 1993.

[2] D. N. Klyshko. Polarization of light: fourth-order effects and polarization-squeezed states. *J. Exp. Theor. Phys.*, 84:1065–1079, 1997.

[3] T. Radhakrishnan. The dispersion, birefringence and optical activity of quartz. *Proc. Indian Acad. Sci. A*, 25:260–265, 1947.

[4] W. A. Shurcliff. *Polarized light*. Harward University Press, 1962.

6 Geometric phase

In this chapter, we discuss a phenomenon that appears in different fields of physics but is most important for polarization optics, where it is known as the Pancharatnam phase. Generally, this effect appears wherever we deal with the trajectory of a point on a curved surface. It has the general term 'topological phase' or 'geometric phase'—the latter will be used throughout this chapter. In quantum physics, this phase is usually called the Berry phase. Below, we start from several simple examples [3].

6.1 Examples of geometric phase

6.1.1 The Foucault pendulum

Imagine a very large pendulum, called the Foucault pendulum (after Leon Foucault who first introduced it), which is used to demonstrate the rotation of the Earth. Such a pendulum can be found in many universities, science museums and other public places like the Panthéon in Paris. During a demonstration, a guide usually starts its swinging motion and marks the oscillation plane. After the pendulum swings for several minutes, the spectators see that the oscillation plane has rotated, due to the fact that the Earth has turned while the plane of the pendulum was constant. A naïve expectation would be that in 24 hours it will return to the initial position. But this will be only the case on the North and South Poles, where an observer will see the plane of oscillations rotating with the same angular speed as the Earth. If the pendulum swings at a point with the latitude y, after 24 hours the plane of rotation will turn by an angle

$$\beta = 2\pi(1 - |\sin y|). \tag{6.1}$$

This effect is illustrated by Fig. 6.1. Here, we can imagine that the Earth is stationary, but the pendulum is displaced around it, along a single parallel, so that its plane of oscillations is constant. In the end it returns to the same point, but the line of oscillations will have to tilt in order to remain tangent to the Earth. In other words, for a stationary observer, the pendulum's plane of oscillations will rotate more slowly than the Earth, and the difference (6.1) will be seen after 24 hours.

At the North or South Pole, where $y = \pm\pi/2$, the Foucault effect is maximal, and the naïve picture will be correct: the plane of the pendulum oscillations will rotate for the observer with the same angular velocity as the Earth and it will recover in 24 hours. But at the equator, $y = 0$, an observer will not see any effect of the Earth rotation, and the angle (6.1) will be 2π. One can notice that

$$\beta = \Omega, \tag{6.2}$$

which is the solid angle that would be covered if the pendulum were driven around the Earth along a parallel (Fig. 6.1).

https://doi.org/10.1515/9783110668025-006

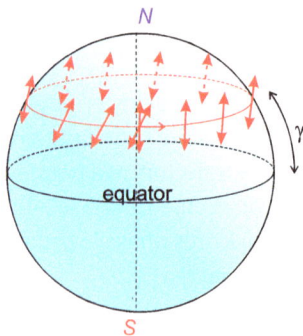

Figure 6.1: Rotation of the plane of the Foucault pendulum.

In another hypothetical example, the Foucault pendulum is indeed driven around the Earth. For instance, let us start such a pendulum at the North Pole, drive it down to the equator along some meridian, coinciding with the direction of swinging, then drive it along the equator by an angle $\pi/2$ and return it back to the North Pole. Figure 6.2 shows the trajectory of the pendulum in orange colour.[1] To understand what happens, we have to shift the vector describing the oscillations (red in the figure) so that it is always in the same plane but still tangent to the surface of the Earth. This, however, leads to its rotation, upon return to the North Pole, by an angle $\pi/2$. We notice that once again, this angle coincides with the solid angle covered by the pendulum on the surface of the Earth.

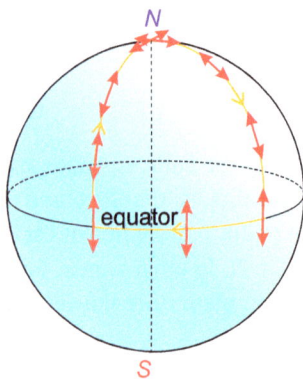

Figure 6.2: A Foucault pendulum driven over the globe, from the North Pole and back. Orange lines show the trajectory of the pendulum and red arrows, the direction of swinging.

This phase (or angular) shift appears because a vector (of the pendulum oscillations) is shifted so that it stays in one plane but yet has to be tangent to a curved space (in this case, the surface of the Earth).

---

1 This situation is only possible if the pendulum has very low damping, so that it is still swinging when it returns.

6.1.2 Non-planar optical path

In another simple example, we consider the polarization of a light beam whose optical path does not lie in a single plane. While in the previous example the oscillation vector of the pendulum had to be tangent to the curved surface of the Earth, now the restriction will be that polarization should be orthogonal to the k-vector and hence to the beam trajectory.

Consider first *the optical tower* (also called a *periscope*), briefly mentioned in Chapter 5. In a periscope, the beam, initially propagating along the x axis and polarized along the z axis (vertically), is reflected by three mirrors, in three mutually orthogonal planes (Fig. 6.3, left panel). After the first mirror, the beam rotates by 90° around the z axis, then mirror 2 rotates it by 90° around the x axis, then mirror 3 rotates it by 90° around the y axis. Eventually the beam is again parallel to the x axis, but now it is horizontally polarized.

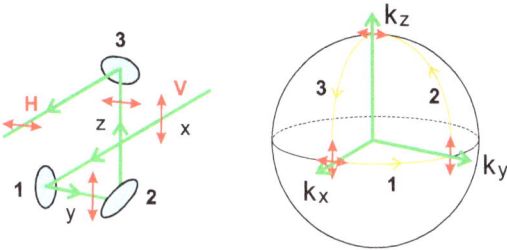

Figure 6.3: The path of a light beam in a periscope (left) and the corresponding trajectory in the k-space (right).

This rotation of polarization can be also related to some solid angle covered in a three-dimensional (3D) space. Indeed, the right panel of Fig. 6.3 shows the evolution of the k-vector of the beam. Initially along the x axis, its trajectory covers a solid angle $\pi/2$ in the k-space and then returns again to the same direction. The 'trajectory' is shown by orange and the electric field direction, by red arrows. The situation is very similar to the one with the Foucault pendulum (Fig. 6.2). As a result, the angle of polarization rotation is again given by the solid angle covered in the 3D space (on a sphere); see Eq. (6.2).

An optical tower is used in laboratories to implement a $\pi/2$ polarization rotation without the use of dispersive elements. At the same time, this effect of polarization rotation can be a problem in experiments where polarization should be maintained to a very high accuracy. Indeed, any reflections of the beam that are not within a single plane will change its polarization state. The change will be the larger, the larger the angle covered by the beam in the 3D space.

Propagation in a fiber. This situation with reflections in different planes can be generalized to the case of light propagating in an optical fiber. If the fiber lies in one plane, the polarization will not be changed. But if the fiber direction forms a trajectory in 3D space, then the polarization will be rotated by an angle given by the solid angle covered by this trajectory. This effect is used in a fiber-based device called 'polarization controller', which will be discussed in Chapter 9, although with a different interpretation.

6.2 Interference of arbitrarily polarized beams

After these simple examples, let us pass to the geometric phase effects caused by polarization transformations. In what follows, we will mainly use the approach developed by Shivaramakrishnan Pancharatnam [6].

In Chapter 1, it was mentioned that orthogonally polarized beams do not form an interference pattern. But what can we say about non-orthogonally polarized beams? To answer this question, consider arbitrary points A, B on the Poincaré sphere (Fig. 6.4): if they are not opposite to each other, they depict polarization states that are not orthogonal. Then, if we try to observe interference between the corresponding beams A, B, the visibility will be nonzero. What value will it have?

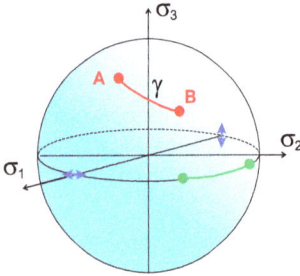

Figure 6.4: Two arbitrary points A, B on the Poincaré sphere and their angular distance γ. Green points: linearly polarized states with the same angular distance on the sphere.

Let the Jones vectors of the two beams be $\vec{e}_A$, $\vec{e}_B$. If beams A and B are overlapped, and they have the same intensity I_0, the total intensity will be given by

$$I = I_0|\vec{e}_A + \vec{e}_B|^2 = 2I_0(1 + |\vec{e}_A \cdot \vec{e}_B| \cos \Delta), \tag{6.3}$$

where Δ is the phase of the interference (as defined by Pancharatnam, and hereafter called the *Pancharatnam phase*). Clearly, the visibility, defined as

$$V \equiv \frac{I_{\max} - I_{\min}}{I_{\max} + I_{\min}}, \tag{6.4}$$

is equal to $|\vec{e}_A \cdot \vec{e}_B|$. (Note that, as mentioned in Chapter 3, the scalar product of two Jones vectors implies the complex conjugation of one of them.) In other words, *the interference visibility for two polarized beams of equal intensity is given by the absolute value of the scalar product of their Jones vectors.*

The absolute value of this scalar product has a clear interpretation on the Poincaré sphere. Indeed, if A and B correspond to linearly polarized states (green points in Fig. 6.4), with coordinates $\vartheta_{A,B}$ (for linearly polarized states, $\varphi_{A,B} = 0$), then the scalar product is [see Eq. (3.6)]

$$|\vec{e}_A \cdot \vec{e}_B| = \cos\frac{\vartheta_A}{2}\cos\frac{\vartheta_B}{2} + \sin\frac{\vartheta_A}{2}\sin\frac{\vartheta_B}{2} = \cos\frac{\vartheta_A - \vartheta_B}{2}. \tag{6.5}$$

But $\vartheta_A - \vartheta_B = \gamma$, the length of the arc between the two green points in Fig. 6.4. (Note that two points on a sphere can be connected by infinitely many arcs; this one is the shortest, called *the geodesic line*.) But because it is only the relative position of two points on the Poincaré sphere that matters, this relation can be generalized: for any two points A, B with the angular separation γ (red points), the visibility of interference will be given by the cosine of half the angular distance between them,

$$V = \cos\frac{\gamma}{2}. \tag{6.6}$$

For instance, the visibility will be zero for opposite points, in full accordance with the statement that orthogonally polarized beams do not form an interference pattern. For points separated by a quadrant, the visibility will be equal to $1/\sqrt{2}$. This will be the case, for instance, for beams polarized vertically and diagonally, or linearly and circularly.

6.3 Decomposition of a beam in two differently polarized components

6.3.1 Decomposition of a beam in two orthogonally polarized components

We saw that the angular length of the geodesic arc between any two points A, B on the Poincaré sphere determines the modulus of the Jones vectors scalar product: $|\vec{e}_A \cdot \vec{e}_B| = \cos\frac{\gamma}{2}$. Suppose we need to decompose an arbitrarily polarized beam, given by a point C on the Poincaré sphere, with the Jones vector $\vec{e}_C$, in two orthogonally polarized beams, given by points A and A' (Fig. 6.5). This is always possible, and the projections of the Jones vector $\vec{e}_C$ on the Jones vectors of the two beams will be

$$|\vec{e}_C \cdot \vec{e}_A| = \cos\frac{\alpha}{2} \tag{6.7}$$

and

$$|\vec{e}_C \cdot \vec{e}_{A'}| = \cos\frac{\pi - \alpha}{2} = \sin\frac{\alpha}{2}. \tag{6.8}$$

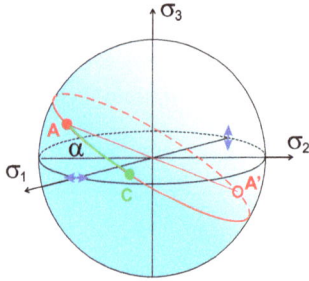

Figure 6.5: Decomposition of a beam with polarization C in two beams with orthogonal polarizations A, A'.

The decomposition will have the form

$$\vec{e}_C = \cos\frac{\alpha}{2}\vec{e}_A + \sin\frac{\alpha}{2}\vec{e}_{A'}. \tag{6.9}$$

In an experiment, this decomposition can be performed with a polarization prism followed by two identical polarization transformation elements. As described in the previous chapter, a combination of the three waveplates QWP + HWP + QWP provides an arbitrary transformation. However, because the beams after a prism are linearly polarized, it is sufficient to have only a HWP and a QWP in each beam after the prism.

6.3.2 Decomposition of a beam in two non-orthogonally polarized components

Similarly, a beam in a polarization state C can be decomposed into two beams in polarization states A and B, the angular distance between them being γ (Fig. 6.6). One can show that the relation between the Jones vectors will be [6]

$$\sin\frac{\gamma}{2}\vec{e}_C = \sin\frac{\beta}{2}\vec{e}_A + \sin\frac{\alpha}{2}\vec{e}_B, \tag{6.10}$$

where α and β are the angular distances from point C to points A and B, respectively.

This relation can be understood by realizing its similarity to the decomposion of a usual vector in a 2D Cartesian space in two non-orthogonal components (Fig. 6.7). If $\vec{A}, \vec{B}, \vec{C}$ are unit vectors, then for the non-orthogonal projection a of $\vec{C}$ on $\vec{A}$ (shown by

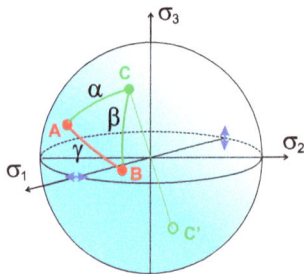

Figure 6.6: Decomposition of a beam with the polarization C in two beams with arbitrary polarizations A, B.

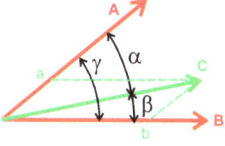

Figure 6.7: Decomposition of a vector in two non-orthogonal vectors.

green dashed line) we have

$$a \sin \gamma = \sin \beta, \tag{6.11}$$

and for the non-orthogonal projection b of $\vec{C}$ on $\vec{B}$ (another green dashed line) we have a similar relation,

$$b \sin \gamma = \sin \alpha. \tag{6.12}$$

Then, for the vectors in Fig. 6.7, we can write

$$\vec{C} = \frac{\sin \beta}{\sin \gamma} \vec{A} + \frac{\sin \alpha}{\sin \gamma} \vec{B}. \tag{6.13}$$

This decomposition almost perfectly coincides with Eq. (6.10). The only difference is that the latter contains halved angles. But this is because on the Poincaré sphere all angles are a factor of two larger than their counterparts in the Cartesian space.

6.4 Pancharatnam phase

6.4.1 Calculation of the Pancharatnam phase

We can now find the Pancharatnam phase according to its definition: if two beams, originally in polarization states A and B, are brought together into a beam with polarization C, the Pancharatnam phase is the phase of their interference. Taking the squared modulus of Eq. (6.10), we get

$$\sin^2 \frac{\gamma}{2} = \sin^2 \frac{\alpha}{2} + \sin^2 \frac{\beta}{2} + 2|\vec{e}_A \cdot \vec{e}_B| \cos \Delta \sin \frac{\alpha}{2} \sin \frac{\beta}{2}, \tag{6.14}$$

or, after substituting the value of the scalar product $|\vec{e}_A \cdot \vec{e}_B|$,

$$\sin^2 \frac{\gamma}{2} = \sin^2 \frac{\alpha}{2} + \sin^2 \frac{\beta}{2} + 2 \cos \frac{\gamma}{2} \cos \Delta \sin \frac{\alpha}{2} \sin \frac{\beta}{2}, \tag{6.15}$$

Then the Pancharatnam phase is given by the relation

$$\cos \Delta = \frac{\sin^2 \frac{\gamma}{2} - \sin^2 \frac{\alpha}{2} - \sin^2 \frac{\beta}{2}}{2 \cos \frac{\gamma}{2} \sin \frac{\alpha}{2} \sin \frac{\beta}{2}}. \tag{6.16}$$

It is convenient to consider, instead of point C, another point C' (Fig. 6.6), opposite to C. For this opposite point, the angular distance to point A is $\alpha' = \pi - \alpha$ and the angular distance to point B, $\beta' = \pi - \beta$. Then Eq. (6.16) can be rewritten as

$$\cos(\pi - \Delta) = \frac{\cos^2 \frac{\alpha'}{2} + \cos^2 \frac{\beta'}{2} + \cos^2 \frac{\gamma}{2} - 1}{2 \cos \frac{\gamma}{2} \cos \frac{\alpha'}{2} \cos \frac{\beta'}{2}}. \tag{6.17}$$

The expression in the right-hand side, according to equations of solid geometry, is equal to the cosine of half the solid angle subtended by the geodesic triangle C'BA on the sphere:

$$\cos(\pi - \Delta) = \cos \frac{\Omega'}{2}. \tag{6.18}$$

Meanwhile, $\Omega' = 2\pi - \Omega$, where Ω is the solid angle subtended by the geodesic triangle ABC on the Poincaré sphere. It follows that

$$\Delta = \frac{\Omega}{2}. \tag{6.19}$$

It means that, *if two beams in polarization states A and B form an interference pattern in the polarization state C, the phase of the interference will be given by half of the solid angle subtended by the geodesic triangle ABC.* One can implement this situation by 'inverting' the scheme we have considered before, with a polarization prism followed by two arbitrary polarization transformations.

In particular, if the interference of states A and B leads to a state C lying on the geodesic connecting them, the solid angle is zero. Correspondingly, the phase of the interference is zero, i. e., the interference is constructive.

This means that, if a polarization state is transformed in such a way that the corresponding point moves along the geodesic line on the Poincaré sphere, no phase shift appears until a half-circle is made. But if the point is moved along a closed contour ABC in Fig. 6.6, then there appears a phase shift: the Pancharatnam phase (6.19).

6.4.2 Measurement of the Pancharatnam phase

Consider now manifestations of the Pancharatnam phase in the experiment. A simple way to observe it is to make the point on the Poincaré sphere go a full circle around the equator [3]. This transformation can be performed by means of a rotator with $\delta = \pi$. The trajectory will subtend a solid angle $\Omega = 2\pi$ on the Poincaré sphere. If the initial polarization state is vertically polarized, then in the course of polarization rotation it becomes horizontal, then again returns to the vertical polarization, but the field vector changes the direction to opposite. Figure 6.8 shows the evolution of the polarization

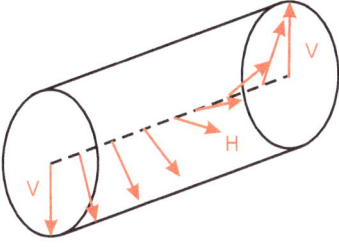

Figure 6.8: Evolution of the field vector in a rotator with a phase $\delta = \pi$. The incident beam is polarized vertically, and so is the output beam, but the phase changes by π.

direction (the electric field vector) along the rotator, which could be, for instance, a cuvette with sugar solution.

Because we ignore the total phase of the Jones vector, we say that the polarization state did not change: light is still vertically polarized. But then the direction of the electric field vector is related to a phase that has been acquired, and if we make this beam interfere with the initial beam, a phase shift of π will appear, which is exactly equal to $\Omega/2$.

Actually we already came across a similar effect in Chapter 5 when we considered the action of a HWP on linearly polarized light. While obtaining Eq. (5.13), we omitted the phase factor i, saying that it was irrelevant for the Jones vector and the final state was linearly polarized. However, if the transformation were performed with a rotator, the trajectory on the Poincaré sphere would be along a geodesic (equator), and no phase factor would emerge. In an experiment, one could see this phase difference by placing a HWP in one arm of an interferometer and a rotator in the other arm.

A simpler experiment is to put HWPs into both arms of a Mach–Zehnder interferometer fed with circularly polarized light. For instance, let the input beam be right-hand circularly polarized. Both plates will transform this state into left-hand circularly polarized light, and at the output there will be perfect (with 100 % visibility) interference pattern. But the phase of the interference will depend on the orientation of the plates. If one of them is oriented with the optic axis horizontal, it performs the rotation on the Poincaré sphere around the σ_1 axis (see Section 5.3). At the same time, the other HWP, oriented at an angle χ, will rotate the initial point around an axis that is in the equatorial plane at an angle 2χ to σ_1. The two trajectories on the Poincaré sphere (shown in Fig. 6.9 by green and magenta colours) will subtend a solid angle $\Omega = 2\chi$. Then the interference phase at the output will be given by the orientation of the second HWP: $\Delta = \chi$. This is the simplest experiment in which the Pancharatnam phase can be directly observed.

This example shows that, from the viewpoint of the Pancharatnam phase, important are not only the initial and final polarization states, but also the trajectory on the Poincaré sphere along which the polarization transformation happens. As mentioned in Chapter 5, there are infinitely many ways to transform the polarization from one state into another, and all these ways are accompanied by different Pancharatnam phase shifts. In particular, as we have just seen, if the rotation is along a geodesic (the

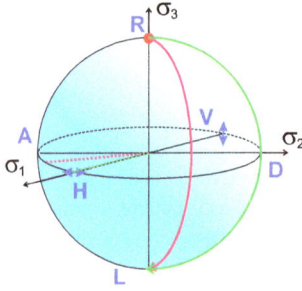

Figure 6.9: Transformations performed by two differently oriented HWPs (horizontally, with green colour, and at an arbitrary angle, with magenta) on the Poincaré sphere. The axes of rotations are shown by dashed lines.

equator, in particular), no phase shift occurs unless a semi-circle is covered. For this reason, if the rotation of linear polarization is performed with a rotator (for instance, polarization is transformed from horizontal to diagonal), no phase shift occurs. Meanwhile, when the same result is achieved with a HWP oriented at $\chi = \pi/8$, then a certain solid angle is covered (the trajectory is closed by completing it with a geodesic line).

Concluding this section, let us stress again that there are two equivalent definitions of the Pancharatnam phase. On the one hand, it is a phase acquired due to the evolution of a point on the Poincaré sphere along a closed trajectory. On the other hand, it is the interference phase for two beams whose polarization states are described by two points coming together along two different parts of the same trajectory.

6.5 Berry phase

The Pancharatnam phase and the geometric phase we considered in the first section of the chapter, although related to different phenomena, have several important features in common. In particular:
1. The phase does not depend on the trajectory, but only on the surface subtended by it.
2. The phase appears due to the constraint imposed on the parallel transport of a vector on a sphere.

In this section, let us briefly discuss other manifestations of the geometric phase, relating to quantum physics. The general concept is known as the *Berry phase* as it has been first described by Michael Berry. While the rigorous consideration can be found in the original work [2], here we will consider only two examples of the Berry phase.

One example is the evolution of a two-level quantum system like a spin 1/2 particle in a magnetic field or a two-level atom interacting with resonant radiation. (Another quantum system with the same description is a polarized single photon, considered in detail in Chapter 11.) The state of a two-level quantum system can be described by the *Bloch vector*, which is introduced in terms of the density matrix ρ_{ij}, $i, j = 1, 2$ [5]:

$$\vec{R} \equiv (2\,\mathfrak{Re}(\rho_{21}), 2\,\mathfrak{Im}(\rho_{21}), \rho_{11} - \rho_{22}). \tag{6.20}$$

For a pure quantum system, the state can be described by a vector $|\Psi\rangle = \alpha|1\rangle + \beta|2\rangle$, where $|1\rangle$ and $|2\rangle$ are the ground and excited states, respectively, and the complex numbers α, β satisfy the normalization condition $|\alpha|^2 + |\beta|^2 = 1$. The density matrix elements are then $\rho_{11} = |\alpha|^2, \rho_{22} = |\beta|^2, \rho_{12} = \alpha\beta^* = \rho_{21}^*$. In this case, $|\vec{R}| = 1$, the Bloch vector is a unit vector, and both its structure and properties are equivalent to the ones of the normalized Stokes vector $\vec{\sigma}$ for polarized light (section 3.3). In full accordance with this analogy, α and β correspond to the components of the Jones vector.

Similarly to the normalized Stokes vector, the Bloch vector is usually depicted as a point on a unit sphere (the *Bloch sphere*), which is therefore equivalent to the Poincaré sphere. The South Pole of the Bloch sphere corresponds to the ground state of the quantum system, and the North Pole, to the excited state. Points on the equator correspond to the system being in a coherent superposition of ground and excited states.

For a mixed state, the Bloch vector has absolute value $|R| < 1$, and this situation is similar to the one of partially polarized light. In this case, the point depicting the state of the two-level system is inside the Bloch sphere.

Similarly to how the Stokes vector rotates due to the polarization transformations with phase plates or polarization rotators, rotation of the Bloch vector describes the evolution of the two-level system under the action of an external force. For instance, a two-level atom driven by an external resonant field performs transitions from the ground state into the excited state and back (Rabi oscillations). Its Bloch vector makes circles along a meridian of the Bloch sphere. If the field is non-resonant, the atom, initially in the ground state, never gets into the excited state, and its Bloch vector makes smaller circles around a point on the equator [5]. These rotations are perfectly similar to the SO(3) rotations of the Stokes vector. And, quite similarly, if such a transformation or a series of transformations covers a closed trajectory, a geometric phase appears, given by the solid angle subtended by the trajectory. This geometric phase acquired by the quantum state of an atom is therefore equivalent to the Pancharatnam phase of polarized light.

An equivalent case is the evolution of a spin 1/2 particle in the alternating magnetic field. The state of the particle is also shown as a point on a Bloch sphere, and variation of the magnetic field leads to the transport of the point over the sphere. A closed trajectory is again associated with the phase shift [2, 3].

Another manifestation of the Berry phase is the Aharonov–Bohm effect [1, 2, 4]. This effect entails that an electron moving along a closed trajectory around a solenoid with magnetic field acquires a phase determined by the magnetic flux through the surface subtended by this trajectory. Similar to the case of the evolution over the Bloch sphere or Poincaré sphere, the phase does not depend on the trajectory itself but only on the area subtended by it. Note that along the trajectory of the electron, the magnetic field can be zero. Alternatively, and similar to the case of the Pancharatnam phase, the geometric phase shift can be observed by making electrons move along two different trajectories forming a closed circuit [4]. At the output, interference can be observed, with the phase scaling as the magnetic field flux through the circuit.

This effect is used in high-precision magnetometers based on Josephson junctions [4]: if two junctions are connected in parallel, they form an interferometer, so that the resulting superconducting current can be enhanced or suppressed due to the interference. If there is a magnetic field between these two junctions, the phase of the interference is determined by the magnetic field flux through the contour formed by the junctions. This way the magnetic flux can be measured with a very high precision.

Bibliography

[1] Y. Aharonov and D. Bohm. Significance of electromagnetic potentials in the quantum theory. *Phys. Rev.*, 115:485–491, 1959.
[2] M. V. Berry. Quantal phase factors accompanying adiabatic changes. *Proc. R. Soc. Lond. A*, 392:45–57, 1984.
[3] D. N. Klyshko. Berry geometric phase in oscillatory processes. *Phys. Usp.*, 36(11):1005–1019, nov 1993.
[4] D. N. Klyshko. Basic quantum mechanical concepts from the operational viewpoint. *Phys. Usp.*, 41(9):885–922, 1998.
[5] D. Klyshko. *Physical foundations of quantum electronics*. World Scientific, 2011.
[6] S. Pancharatnam. Generalized theory of interference and its applications. *Proc. Indian Acad. Sci.*, 44:247, 1956.

7 Structured light

So far, we considered only the polarization state of propagating plane waves. In the general discussion, we conveniently referred to light beams without introducing a theoretical framework beyond plane waves. However, to better understand the spatial degrees of freedom for light fields, we now need to turn away from single plane waves and consider more realistic solutions to Maxwell's equations. In this context we will also learn that light, although usually described as a transverse electromagnetic wave, cannot be described in full compliance with Maxwell's equations if only transverse field (polarization) components orthogonal to the mean propagation direction are assumed. We need to take into account also longitudinal field components (see discussion in Chapter 8), i. e., electric and magnetic field components oscillating along the direction of propagation. Additionally and unexpectedly, we will also show that the spatial distribution of light propagating in free space can even feature points or lines where only electric or magnetic fields are present (similar to standing waves). The appearance of such purely electric or magnetic fields is intimately connected with the aforementioned longitudinal components. But here we start our discussion with the derivation of the paraxial wave or Helmholtz equation resulting in analytical and approximate beam solutions. Afterwards, we discuss the spatial structure of light beams from a scalar and a vectorial perspective. At the end of the chapter, different methods for the generation of structured light will be summarized.

7.1 The paraxial wave equation

We start again with Maxwell's universal equations (2.7) discussed in Chapter 2 to derive the full wave equation. For the sake of simplicity, we assume propagation in vacuum (non-magnetic medium with no charges or currents present). We apply $\vec{\nabla} \times$ to the curl equation for the electric field [26], resulting in

$$\vec{\nabla} \times \vec{\nabla} \times \vec{E} = -\mu_0 \vec{\nabla} \times \dot{\vec{H}}. \tag{7.1}$$

We omit the time dependence here for brevity. If we assume a harmonic time dependence of the fields ($\propto e^{-i\omega t}$), we get

$$\vec{\nabla} \times \vec{\nabla} \times \vec{E} = i\omega\mu_0 \vec{\nabla} \times \vec{H}. \tag{7.2}$$

We now substitute the term $\vec{\nabla} \times \vec{H}$ with the curl equation for the magnetic field,

$$\vec{\nabla} \times \vec{\nabla} \times \vec{E} = -\omega^2 \mu_0 \epsilon_0 \vec{E}. \tag{7.3}$$

This equation can be further simplified by taking advantage of the vector algebra rule (4.23)

$$\vec{\nabla} \times \vec{\nabla} \times \vec{E} = \vec{\nabla}(\vec{\nabla} \cdot \vec{E}) - \vec{\nabla}^2 \vec{E}. \tag{7.4}$$

https://doi.org/10.1515/9783110668025-007

Together with $\vec{\nabla} \cdot \vec{E} = 0$, $\mu_0 \epsilon_0 = 1/c^2$ and $k = \omega/c$, we finally end up with the (full) wave or Helmholtz equation:

$$(\vec{\nabla}^2 + k^2)\vec{E} = 0. \tag{7.5}$$

In an equivalent manner, the corresponding equation for the magnetic field can be derived.

In this chapter, we are interested in the retrieval of analytical solutions representing light beams propagating in a paraxial and, hence, a collimated fashion, just like light beams emitted by a laser. This restriction will allow us to derive a paraxial wave equation. Without loss of generality, we let the light propagate along the z-direction. Hence, the electric field is assumed to depend on the coordinate z as $\vec{E}(x,y,z) = \vec{E}_0(x,y,z)e^{ikz}$. If we substitute the electric field in (7.5) with this dependence, we get

$$\left(\frac{\partial^2}{\partial x^2} + \frac{\partial^2}{\partial y^2} + \frac{\partial^2}{\partial z^2} + k^2 \right)\vec{E}_0(x,y,z)e^{ikz} = 0, \tag{7.6}$$

leading, after several steps, to the following equation:

$$\left(\frac{\partial^2}{\partial x^2} + \frac{\partial^2}{\partial y^2} \right)\vec{E}_0(x,y,z) + \frac{\partial^2}{\partial z^2}\vec{E}_0(x,y,z) + 2ik\frac{\partial}{\partial z}\vec{E}_0(x,y,z) = 0. \tag{7.7}$$

If we assume propagation in a paraxial and collimated fashion, i. e., we expect the intensity of the beam on the optical axis to vary only very slowly with propagation (along z) in comparison to its variation in the transverse direction (x, y), the term with the second derivative in z can be neglected. We obtain

$$\left(\vec{\nabla}_\perp^2 + 2ik\frac{\partial}{\partial z} \right)\vec{E}_0(x,y,z) = 0, \tag{7.8}$$

with $\vec{\nabla}_\perp^2 = \frac{\partial^2}{\partial x^2} + \frac{\partial^2}{\partial y^2}$. Equation (7.8) is usually referred to as the paraxial wave or Helmholtz equation. This equation can be solved analytically. In the next section, selected scalar solutions to this equation will be introduced and discussed. It should be noted here already that solutions to the paraxial wave equation do not satisfy Maxwell's equations or the full wave equation, because they resulted from an approximation. This will become crucial in the discussion of three-dimensional fields in Chapter 8. However, solutions to Eq. (7.8) represent very good approximations for propagating collimated light beams we use in the lab. In the following section we discuss some of the most prominent solutions of Eq. (7.8).

7.2 Structured scalar light beams—transverse phase patterns and phase singularities

Before discussing light beams tailored spatially in their polarization, we start with a class of light beams usually referred to as scalar spatial modes. These are light beams featuring a polarization state that is fixed and homogeneous across their lateral extent and upon propagation, while the intensity (and phase) are structured and, hence, non-homogeneous. These solutions can be treated in a scalar framework, i. e., the polarization state does not depend on the spatial coordinates. Several sets of scalar spatial modes can be retrieved from the paraxial wave equation (7.8). Depending on the coordinate system (Cartesian, cylindrical, etc.), or more practically speaking, in dependence on the symmetry of the cavity mirrors in a laser from which these modes originate, various complete and orthogonal sets of spatial scalar modes can be derived, including, but not limited to, Hermite–Gaussian (HG) or Laguerre–Gaussian (LG) modes. These modes earn their names from the Hermite or Laguerre polynomials they depend on, which are multiplied by a Gaussian function. The fundamental mode $\vec{E} = \vec{E}_{0,0}^G$ of both LG and HG mode sets is the familiar Gaussian light beam,

$$\vec{E}_{0,0}^G(x,y,z) = \vec{E}^G \frac{1}{1+(2iz/kw_0^2)} e^{-\frac{x^2+y^2}{w_0^2[1+(2iz/kw_0^2)]}} e^{ikz}, \tag{7.9}$$

with w_0 denoting the beam waist (radius at $z = 0$) and $\vec{E}^G$ the actual position-independent transverse polarization of the beam. This analytical expression for a paraxially propagating Gaussian beam can be rewritten in the following way:

$$\vec{E}_{0,0}^G(x,y,z) = \vec{E}^G \frac{w_0}{w(z)} e^{-\frac{x^2+y^2}{w^2(z)}} e^{i[kz+\frac{k(x^2+y^2)}{2R(z)}-\eta(z)]}. \tag{7.10}$$

The beam radius $w(z)$, the wave front radius of curvature $R(z)$, and the Gouy phase term $\eta(z)$ [18] are defined as follows:

$$w(z) = w_0 \sqrt{1 + \frac{z^2}{z_R^2}}, \tag{7.11}$$

$$R(z) = z\left(1 + \frac{z_R^2}{z^2}\right), \tag{7.12}$$

$$\eta(z) = \arctan\left(\frac{z}{z_R}\right), \tag{7.13}$$

with $z_R = \frac{kw_0^2}{2}$ being the Rayleigh range (see Fig. 7.1).

As we can see, the polarization as well as the beam shape can be approximated to be constant upon propagation along the z-axis, while the beam diameter, the radius of phase-front curvature, as well as an additional phase term (η) are z-dependent. The Gouy phase term defines the relative phase lag between a Gaussian beam propagating from $-\infty$ to $+\infty$ and a plane wave. It reflects the converging and diverging nature of beam propagation.

The aforementioned higher-order HG modes (in a Cartesian coordinate frame) form a complete set of modes and can be constructed from the fundamental Gaussian

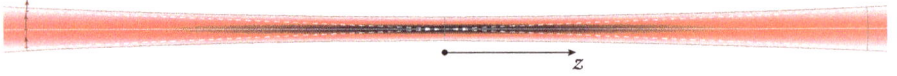

Figure 7.1: Paraxial propagation of a beam of light. The beam with a Gaussian intensity profile propagates along the horizontal (z) axis and reaches its smallest radius (beam waist w_0) in the center of the sketch where the phase front is planar and the radius of the phase-front curvature diverges. The dashed white lines indicate the geometrical rays crossing at the focus and defining the angular spread of the beam for large distances from the waist plane. The intensity along the optical axis changes only very slowly (small convergence and divergence angle) allowing for the application of a paraxial approximation.

solution in the following manner:

$$\vec{E}^{HG}_{m,n}(x,y,z) = w_0^{m+n} \frac{\partial^m}{\partial x^m} \frac{\partial^n}{\partial y^n} \vec{E}^G_{00}(x,y,z), \tag{7.14}$$

with m, n non-negative integer indices. This construction is based on the inherent properties of Hermite polynomials. The full set of Hermite–Gaussian modes can therefore be summarized with the following field expression:

$$\vec{E}^{HG}_{m,n}(x,y,z) = \vec{E}^{HG} \frac{w_0}{w(z)} H_m\left(\sqrt{2}\frac{x}{w(z)}\right) H_n\left(\sqrt{2}\frac{y}{w(z)}\right)$$
$$\cdot e^{-\frac{x^2+y^2}{w^2(z)}} e^{i[kz + \frac{k(x^2+y^2)}{2R(z)} - \eta_{HG}(z)]}, \tag{7.15}$$

with $\eta_{HG}(z)$ being the generalized Gouy phase term defined by $\eta_{HG}(z) = (m + n + 1)\arctan(\frac{z}{z_R})$. The polynomial pre-factors result in changes of the phase within the beam cross-section (beam profile for a fixed z position), affecting also the intensity. The modes therefore naturally feature a structured (transversely varying) phase distribution in combination with a non-homogeneous intensity pattern. In addition, Hermite functions are invariant under the Fourier transform and, hence, keep their shapes upon propagation. Selected first-order $(m,n) = (0,1)$ (a) and $(m,n) = (1,0)$ (b) HG modes are shown in Fig. 7.2. As can be seen, the modes feature multiple intensity lobes with their number depending on the chosen indices, while the phase changes from lobe to lobe by π.

Similarly, another prominent and full set of solutions can be derived, i. e., the aforementioned LG modes based on Laguerre polynomials. These modes feature cylindrically symmetric ring-like intensity distributions, with the number and size of rings depending on the chosen indices. The full set of modes can be described by the equations

$$\vec{E}^{LG}_{l,p}(r,\phi,z) = \vec{E}^{LG} \frac{w_0}{w(z)} \left(\frac{r\sqrt{2}}{w(z)}\right)^{|l|} L_p^{|l|}\left(\frac{2r^2}{w(z)}\right).$$
$$\cdot e^{il\phi} e^{-\frac{r^2}{w^2(z)}} e^{i[kz + \frac{kr^2}{2R(z)} - \eta_{LG}(z)]}, \tag{7.16}$$

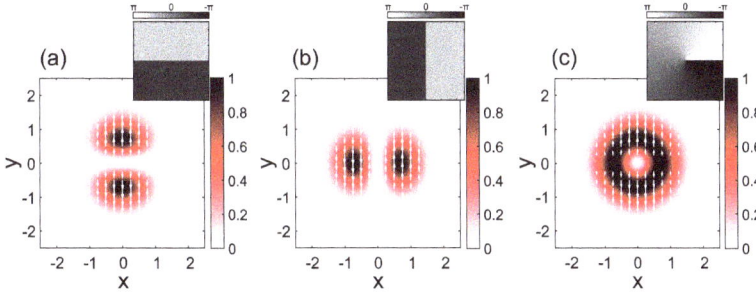

Figure 7.2: Examples of first-order HG (a, b) and first azimuthal order LG (c) paraxial modes, respectively. Distributions of the normalized (electric field) intensity (color-coded), phase (inset) and a snapshot of the electric field vectors (white arrows) are shown. Figure reproduced from [2].

where l is the azimuthal and p the radial index defining the order of the mode and relating to the azimuthal (ϕ) and radial (r) coordinate in the cylindrical coordinate system. $\eta_{\mathrm{LG}}(z)$ is the generalized Gouy phase term defined by $\eta_{\mathrm{LG}}(z) = (|l| + 2p + 1)\arctan(\frac{z}{z_R})$. Figure 7.2(c) also shows intensity and phase distributions of the selected LG mode. As a direct consequence of the term $e^{il\phi}$, LG modes of azimuthal order $l = 1$ ($l = -1$) or higher (lower) exhibit an azimuthally varying phase gradient of $l2\pi$ leading to a spiral phase front. The phase is undefined at the origin (optical axis) where the corresponding modes are dark (intensity reaches zero). Such a point of darkness (or lines of darkness along the beam axis) is referred to as phase singularity or phase vortex. Due to the topological nature of this feature, l is usually also called the topological charge of the vortex. Optical vortices can also be observed when multiple light waves interfere, for instance, in the case of diffraction.

It should also be noted here that LG modes can also be constructed from HG modes and vice versa. This is particularly easy to see for first (azimuthal) order LG modes, which can be constructed by superposing two spatially orthogonal HG modes with indices $(0, 1)$ and $(1, 0)$ polarized along the same axis (see Fig. 7.2(a) and (b)). When superposed in-phase or with a phase delay of π, the resultant mode still features an HG mode shape but appears rotated by ± 45 deg. However, if the two constituent modes are dephased by $\pm\frac{\pi}{2}$, a first ($l = \pm 1$) order LG mode is created featuring a ring-like intensity distribution and the aforementioned spiral phase front (Fig. 7.2(c)). Based on this simple construction it is easy to see that an LG beam of, e. g., first azimuthal order shows a first-order HG modal profile for a fixed snapshot in time. This simple pattern orbits or spins around the optical axis at light's frequency with time evolving. If HG, LG or other paraxial modes of different orders are superposed, interesting propagation-induced effects appear as a direct consequence of mode-order-dependent Gouy phases [3, 9, 31, 44].

In this context it is also worth mentioning that the spiral phase fronts and, therefore, the azimuthal mode indices of LG beams are also associated with another very interesting property of electromagnetic waves, i. e., the orbital angular momentum (OAM) [1, 4, 39]. This property is inherently linked to the spatial structure of a light field and is attributed to the analogy with mechanical systems orbiting around one another. In case of a light beam, the field amplitude distribution orbits about the optical axis while propagating with time passing. This feature finds a variety of different applications, ranging from optical manipulation and trapping to optical communication, quantum key distribution, and beyond [33, 36]. It is also one of light's topological features, which will not be discussed in more detail here. In the next chapter, we come back to a special type of topological property related to the main topic of this book, the polarization of light.

The paraxial wave equation has also many other classes of solutions, including, but not limited to, Bessel–Gaussian (non-diffracting [13, 14]) and hypergeometric-Gaussian (non-orthogonal [24]) modes. For the sake of convenience, we only discussed the most prominent solutions here.

7.3 Vectorial spatial modes and light beams—non-homogeneous polarization distributions

The transition from a single homogeneously polarized plane wave propagating along a given direction (wave vector) to beam-like solutions of the paraxial wave equation indicated already that in a realistic setting, the electromagnetic field must be spatially confined along at least the transverse dimensions (finite beam width), resulting consequently in a spatially varying, or structured intensity or phase distribution. However, we implicitly assumed in this discussion that the light beams feature a homogeneous polarization across the beam profile, which also does not change upon propagation. In other words, we treated the beams in a scalar fashion where the homogeneous polarization could be separated from the spatial profile. In this section, we now extend our discussion to include light's polarization and show that also the polarization is a spatial degree of freedom, resulting in interesting, versatile and widely applicable spatial vectorial modes of light.

The simplest and undoubtedly most prominent class of spatial vectorial light modes are cylindrical vector beams. These modes feature a cylindrical or axial symmetry in their intensity and polarization distribution. For example, radially (and azimuthally) polarized beams feature a locally linear polarization with the polarization plane spanned by the optical axis and the radial (tangential) unit vector, comparable to the spokes of a wheel (see Fig. 7.3). The electric field crosses zero at the beam center (optical axis). Consequently, the intensity profile of cylindrical vector beams resembles the shape of a doughnut and is identical to that of first (azimuthal) order LG modes.

An easy theoretical way to construct a paraxial light beam featuring a cylindrical polarization distribution is similar to the construction of LG modes of first azimuthal

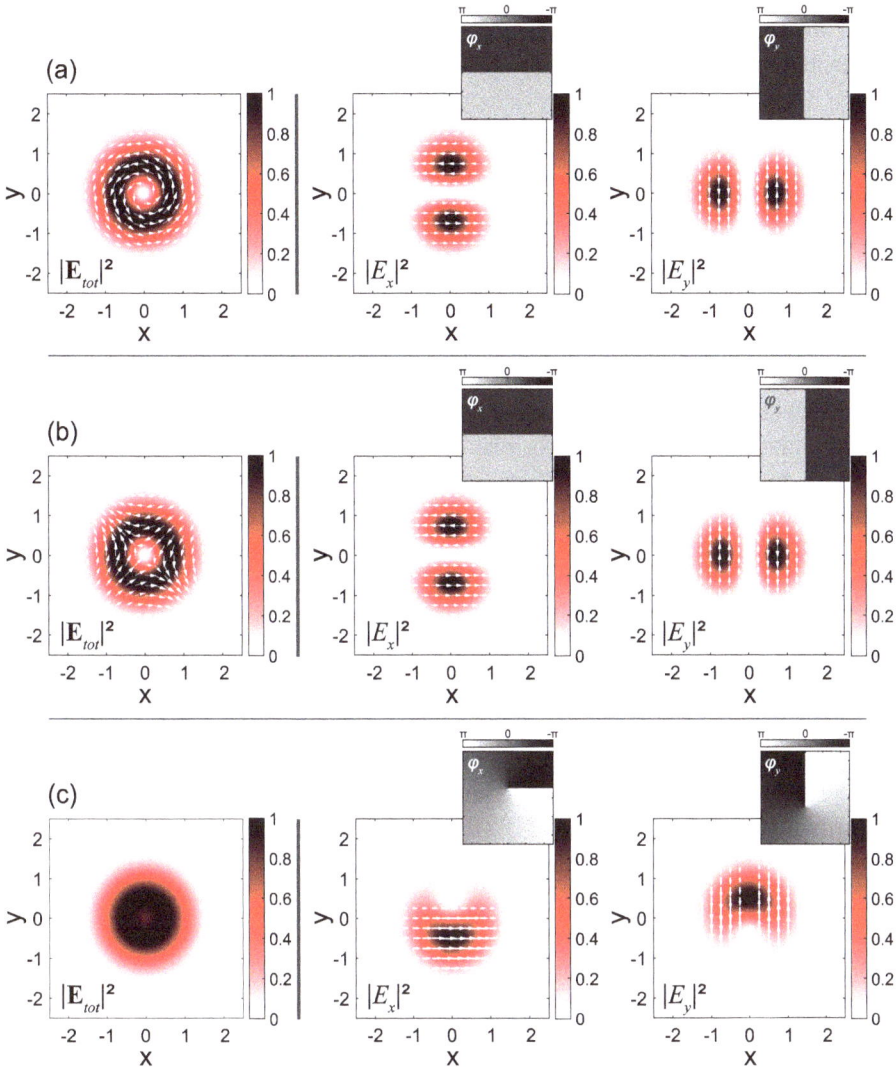

Figure 7.3: Selected examples of paraxial (a) azimuthally and (b) counter-rotating azimuthally polarized as well as (c) Poincaré-type beams. Left: Distributions of the normalized total (electric field) intensity (color-coded). Center and right: Color-coded (electric field) intensity distributions of the transverse electric field components E_x and E_y. Snapshots of the electric field vectors (white arrows) are shown as overlays. For better visibility, the field vectors are not shown for the full electric field intensity of the Poincaré beam in (c). Maps of the relative phase between the field components are shown as insets. Figure reproduced from [2].

order. We can superpose two of the above-mentioned scalar spatial modes, e. g., HG modes, but now with orthogonal polarization states,

$$\vec{E}_{\text{radial}} = E_{1,0}^{HG}\vec{e}_x + E_{0,1}^{HG}\vec{e}_y, \tag{7.17}$$

$$\vec{E}_{\text{azimuthal}} = E_{0,1}^{HG}\vec{e}_x - E_{1,0}^{HG}\vec{e}_y. \tag{7.18}$$

Alternatively, also the superposition of two scalar LG modes of azimuthal index 1 and −1 carrying opposite handedness of circular polarization results in either radial or azimuthal polarization. It should be noted here that depending on the chosen relative phase and polarization of the two constituent modes brought together, also beams carrying a spiral polarization [15] can be created, finding interesting applications especially when highly confined spatially (see the discussion in Chapter 8). These beams can be visualized very instructively when overlapping a radially and an azimuthally polarized doughnut beam, which oscillate either in phase or with a phase-shift of π. Locally, the beam therefore exhibits a non-zero radial and simultaneously azimuthal electric field component, forming a spiral-like pattern of the field for a snapshot in time, imprinting a certain type of *handedness* [15, 21, 41].

Radial, azimuthal and spiral cylindrical vector beams all feature a locally linear electric field, which rotates clockwise when walking around the optical axis in a clockwise sense. These beams therefore have a polarization order of 1, i. e., a rotation of 2π for a full trip around the axis. The polarization order is therefore defined in a similar fashion like before the azimuthal order of LG modes, which itself indicated the number of 2π azimuthal phase changes. Equivalently, also the polarization order can be larger than 1 (in discretized steps of 1) if the number of turns of the linear polarization is more than 2π [25, 40]. If a cylindrical vector beam carries a negative polarization order, the electric field necessarily rotates counter-clockwise.

Although cylindrical vector beams discussed so far feature interesting polarization patterns, they are still linearly polarized everywhere with only the orientation of the polarization plane varying spatially. However, this is by far not the end of the story. Light beams may also feature multiple different polarization types, from linear via elliptical to circular, all present in one cross-section. This important remark directly leads us to the discussion of another very interesting class of spatial vectorial modes of light, which definitely deserve it to be mentioned here also because it is intimately connected to complex polarization patterns (see next section), i. e., so-called *Poincaré* beams [5] (see Fig. 7.3). On the surface of the Poincaré sphere (see Chapter 3), all possible homogeneous polarization states are contained. A Poincaré beam is therefore a beam featuring spatially varying polarization states covering either the full Poincaré sphere (full Poincaré beams) or a part of it. In a certain sense, radially and azimuthally polarized modes can be seen as trivial versions of Poincaré beams, because they cover the full equator. Also this class of beams can be constructed, e. g., by superposing modes we met with already [3, 5]. For instance, if a circularly polarized fundamental Gaussian beam is co-propagating with a first (or minus first) order LG mode of opposite handedness, the resulting beam will be circularly polarized on the optical axis, elliptically polarized off-axis, where both modes feature non-zero intensity, opposite handedness and unequal amplitudes, and linearly polarized, where the two modes have the same amplitude. In addition, the orientation of the polarization ellipse is ruled by the azimuthally changing phase of the LG mode, hence covering a large portion of the Poincaré sphere's surface.

It should be noted here that beams of different order accumulate different phases upon propagation (Gouy phase) influencing the relative polarization in the transverse plane depending on the propagation distance [3, 9, 31, 44].

As we will see in Chapter 8, the spatial confinement of light will naturally lead to an even more complex structure of the electric and/or magnetic components of electromagnetic fields.

7.4 Polarization singularities and generic ellipse fields

The concept of polarization structured beams can also be generalized. Before we move on to the discussion of the spatial degree of freedom of light's vectorial properties in generic fields, we shortly revisit the definition of polarization based on the polarization ellipse (see Chapter 3). We introduced the polarization ellipse as a geometrical concept to describe the state of polarization. It is defined by the orientation and lengths of its major and minor semiaxes a and b. By definition, a and b are *directors* rather than vectors. Berry introduced a convenient and complete definition of the polarization ellipse based on the aforementioned semiaxes together with an additional parameter c, which is oriented normal to the ellipse and defines the handedness of the spinning field (spinning sense) [6]:

$$a = \frac{\Re(\vec{E}\sqrt{\vec{E}^* \cdot \vec{E}^*})}{|\sqrt{\vec{E} \cdot \vec{E}}|}, \tag{7.19}$$

$$b = \frac{\Im(\vec{E}\sqrt{\vec{E}^* \cdot \vec{E}^*})}{|\sqrt{\vec{E} \cdot \vec{E}}|}, \tag{7.20}$$

$$c = \Im(\vec{E}^* \times \vec{E}), \tag{7.21}$$

with the spatial dependence of the field and the ellipse parameters omitted. c is also proportional to the spin density, which defines the local degree of circular polarization. If a and b are of the same length, the polarization ellipse turns into a circle, the field is locally circularly polarized. The orientation of the ellipse is not defined anymore and it does not feature distinguishable major or minor semiaxes. Such a point of circular polarization is therefore usually referred to as polarization singularity in general and a *C-point* in particular. Similarly, if a field is locally linearly polarized, the ellipse is a line and the minor axis b as well as the parameter c are zero, and thus singular as well. Such a point is therefore called an L-point.[1]

The convenient and powerful notation of polarization singularities and polarization distributions in generic 2D fields, following that of scalar phase singularities and their surrounding in scalar fields or beams, was introduced and studied in detail by Nye, Hajnal, Berry, Dennis, Soskin, Freund and others [6, 7, 11, 16, 19, 20, 32, 38]. They also found that around C-points, fields of elliptical polarization form, which take on specific distributions with respect to the ellipse orientations in 2D (see Fig. 7.4). These distributions, together with their polarization singularity in the center, form topological structures, just like the phase singularities and the phase map around them. The polarization ellipses for the most generic fields rotate by $\pm\pi$ when walking along a closed circle around the central C-point, resulting in the definition of a topological index of $\pm\frac{1}{2}$ with the sign depending on the sense of rotation of the ellipses (rotating clockwise or counter-clockwise for a clockwise path). Hence, the ellipses only rotate by 180 degrees in contrast to the polarization order discussed above, which was equal to full integer numbers and, thus, full 360 degree rotation of the electric field vector for one round-trip. This is possible because the major and minor semiaxes of the polarization ellipse are *directors* and not vectors. A rotation by 180 degrees brings us back to the original orientation of the ellipse. The ellipse field is usually surrounded by a closed line of linear polarization separating points of elliptical polarization of opposite handednesses. The topological index can also be higher (single or multiple full or half turns of the ellipse) for more complicated distributions.

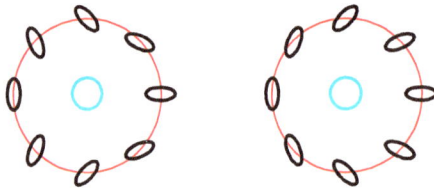

Figure 7.4: Schematic representation of C-points (blue) and generic ellipse fields (black) surrounding them: *star* (left) with topological index $-\frac{1}{2}$ and *lemon* (right) with topological index $+\frac{1}{2}$.

Two fundamental distributions of ellipse fields around C-points are depicted in Fig. 7.4. Based on their shape they are called *star*, *lemon* and *monstar* (*le-mon-star*; not shown here) [7]. The star-type ellipse field distribution is similar to the field in the cross-section of the above-mentioned Poincaré beam with a central C-point surrounded by elliptically polarized light with position-dependent ellipse orientation (see Fig. 7.3). In Chapter 8, we extend our discussion to cover also intriguing phenomena appearing in 3D ellipse fields.

---

1 In 2D ellipse fields, points of linear polarization can usually be found along lines, while C-points are isolated.

7.5 Basic principles of structured light beam generation

Spatial scalar and vectorial modes of light find an ever increasing number of applications. Hence, countless methods for generating different classes of spatial modes have been discussed in the literature. Over the last decades, many more active or passive methods have been proposed, implemented and optimized. Techniques proposed and implemented to date range from local polarization or phase manipulation as well as beam combining, all the way to intra-cavity mode modifications [36, 45]. A detailed and encyclopedic review of such methods goes beyond the scope of this book. However, we want to briefly sketch the basic operational principle of one of the most convenient and powerful fundamental approaches, i. e., the generation of structured paraxial light based on a position-variant polarization or phase manipulation. More details with respect to the corresponding basic building-blocks and transformations acting on polarization and phase of a light beam are discussed in Chapters 5 and 9.

The simplest and most instructive vectorial beam shaper is based on the action of simple waveplates on the polarization of a light beam [12, 22, 27, 34, 37]. To understand the basic idea behind such so-called segmented waveplates, we sketch the following setting. A linearly polarized Gaussian beam of light propagating in a paraxial fashion along the z-axis impinges on an optical element placed in the xy-plane. It consists of 4 (or more) HWPs arranged and cut such that each segment features a different orientation of its fast axis. With the fast axes pointing in different directions in the transverse plane, the polarization of the homogeneously polarized input beam is rotated and, hence, manipulated position-dependently, resulting in a segmented beam of light with differently oriented polarization planes. If the fast axes are chosen appropriately, the resulting polarization pattern in the beam cross-section shows a significant overlap with, e. g., a radially (or azimuthally) polarized beam. However, owing to the limited number of segments of the segmented waveplate, the resulting beam is not in a pure radially (or azimuthally) polarized mode, but it is a superposition of multiple modes. Nonetheless, it can be mode-cleaned by utilizing a cavity or a Fourier (pinhole) filter. Segmented waveplates have been implemented in different configurations [12, 22, 27, 34, 37] naturally showing a quite limited range of applications resulting directly from the inherent wavelength dependence of waveplates.

For acting locally on the (phase and) polarization of an impinging light beam, a variety of different methods have been discussed and studied in detail, involving but not limited to segmented waveplates discussed before, sub-wavelength metallic or dielectric gratings [8, 17], liquid crystal cells [28, 40] or more recently also metasurfaces built from individual plasmonic or dielectric nanoantennas [23, 35, 42, 43]. The field of metasurfaces, which came to life after the inception of effective optical materials or metamaterials, also got boosted by the ever increasing capabilities of modern nanofabrication. For all aforementioned approaches, the field of a light beam to be converted into a spatial mode is manipulated by acting on the beam's local amplitude, phase or polarization. By the implementation of a position-dependent modification of these parameters, the beam can be engineered spatially on demand and almost limitless. The utilization of liquid crystal technology adds an extra level of control to such methods. Liquid crystal molecules show strong optical birefringence thus acting like miniaturized waveplates and phase-shifters (see also Section 4.2.4), while also aligning with an external electric field, which can be used to control the effective birefringence or induced phase precisely. This enables fine control over the beam shape and quality, and it also allows for spectral tunability. The spatial degree of freedom in the corresponding manipulation of polarization and phase is achieved by different means, for instance, by imposing a spatial orientation of the molecules by structured alignment layers or by sub-dividing a liquid crystal cell into pixels and providing pixel-by-pixel voltage control. The latter class of devices is referred to as spatial light modulators (SLM), described in more detail in Chapter 9. They can be implemented as reflective or transmissive devices acting on the local phase of a homogeneously polarized input mode. Position-dependent manipulation is realized via discretized pixel arrays controllable individually. An SLM can therefore be run like a computer screen, turning it into a very versatile, highly flexible and extraordinarily powerful device for beam shaping, imposing almost no limits with respect to the mode order or type to be generated. Despite their phase-only (and amplitude) operation, SLMs were also successfully utilized for the generation of cylindrical vector beams and other vectorial spatial modes [10, 29, 30]. Hence, SLMs are the right choice if flexibility with respect to mode type or order as well as wavelength is required. In addition, so-called q-plates [28], which are also based on voltage-driven liquid crystal cells (see also Section 9.6.2) and position-dependent manipulation of the field in the sense of a HWP, have been established as simple yet highly efficient and reliable devices finding various areas of applications. The underlying idea is the same as the one discussed above for segmented waveplates. The liquid crystal molecules behave like microscopic HWPs and feature azimuthally varying orientations, hence influencing the polarization in a position-dependent manner. In particular, they are usually implemented for the generation of LG modes or cylindrical vector beams. However, they only allow for the generation of a specific group of modes defined by the chosen arrangement of liquid-crystal molecular chains. The concept behind q-plates is discussed in more detail in Section 9.6.2.

Bibliography

[1] L. Allen, M. W. Beijersbergen, R. J C. Spreeuw, and J. P. Woerdman. Orbital angular momentum of light and the transformation of Laguerre–Gaussian laser modes. *Phys. Rev. A*, 45(11):8185, 1992.

[2] P. Banzer. Nano-optics and plasmonics with complex spatial modes of light—structured electromagnetic fields at the nanoscale, 2019. Habilitation Thesis.

[3]   T. Bauer, P. Banzer, E. Karimi, S. Orlov, A. Rubano, L. Marrucci, E. Santamato, R. W. Boyd, and G. Leuchs. Observation of optical polarization Möbius strips. *Science*, 347(6225):964–966, 2015.

[4]   V. Y. Bazhenov, M. S. Soskin, and M. V. Vasnetsov. Screw dislocations in light wavefronts. *J. Mod. Opt.*, 39(5):985–990, 1992.

[5]   A. M. Beckley, T. G. Brown, and M. A. Alonso. Full Poincaré beams. *Opt. Express*, 18(10):10777–10785, 2010.

[6]   M. V. Berry. Index formulae for singular lines of polarization. *J. Opt. A, Pure Appl. Opt.*, 6(7):675, 2004.

[7]   M. V. Berry and J. H. Hannay. Umbilic points on Gaussian random surfaces. *J. Phys. A, Math. Gen.*, 10(11):1809, 1977.

[8]   Z. Bomzon, G. Biener, V. Kleiner, and E. Hasman. Real-time analysis of partially polarized light with a space-variant subwavelength dielectric grating. *Opt. Lett.*, 27(3):188–190, 2002.

[9]   F. Cardano, E. Karimi, L. Marrucci, C. de Lisio, and E. Santamato. Generation and dynamics of optical beams with polarization singularities. *Opt. Express*, 21(7):8815–8820, 2013.

[10]   V. Chille, S. Berg-Johansen, M. Semmler, P. Banzer, A. Aiello, G. Leuchs, and C. Marquardt. Experimental generation of amplitude squeezed vector beams. *Opt. Express*, 24(11):12385–12394, 2016.

[11]   M. R. Dennis. Polarization singularities in paraxial vector fields: morphology and statistics. *Opt. Commun.*, 213(4–6):201–221, 2002.

[12]   R. Dorn, S. Quabis, and G. Leuchs. Sharper focus for a radially polarized light beam. *Phys. Rev. Lett.*, 91(23):233901, 2003.

[13]   J. Durnin. Exact solutions for nondiffracting beams. i. The scalar theory. *JOSA A*, 4(4):651–654, 1987.

[14]   J. Durnin, J. J. Miceli Jr, and J. H. Eberly. Diffraction-free beams. *Phys. Rev. Lett.*, 58(15):1499, 1987.

[15]   J. S. Eismann, M. Neugebauer, and P. Banzer. Exciting a chiral dipole moment in an achiral nanostructure. *Optica*, 5(8):954–959, 2018.

[16]   I. Freund. Polarization flowers. *Opt. Commun.*, 199(1–4):47–63, 2001.

[17]   Z. Ghadyani, I. Harder, N. Lindlein, A. Berger, W. Iff, I. Vartiainen, and M. Kuittinen. Concentric ring metal grating for generating radially polarized light. *Appl. Opt.*, 50:2451, 2011.

[18]   L. G. Gouy. Sur une propriété nouvelle des ondes lumineuses. *C. R. Acad. Sci. Paris*, 110:1251, 1890.

[19]   J. V. Hajnal. Singularities in the transverse fields of electromagnetic waves. I. Theory. *Proc. R. Soc. Lond. Ser. A, Math. Phys. Sci.*, 414(1847):433–446, 1987.

[20]   J. V. Hajnal. Singularities in the transverse fields of electromagnetic waves. II. Observations on the electric field. *Proc. R. Soc. Lond. Ser. A, Math. Phys. Sci.*, 414(1847):447–468, 1987.

[21]   A. Holleczek, A. Aiello, C. Gabriel, C. Marquardt, and G. Leuchs. Classical and quantum properties of cylindrically polarized states of light. *Opt. Express*, 19(10):9714–9736, 2011.

[22]   J. Kalwe, M. Neugebauer, C. Ominde, G. Leuchs, G. Rurimo, and P. Banzer. Exploiting cellophane birefringence to generate radially and azimuthally polarised vector beams. *Eur. J. Phys.*, 36(2):025011, 2015.

[23]   E. Karimi, S. A. Schulz, I. De Leon, H. Qassim, J. Upham, and R. W. Boyd. Generating optical orbital angular momentum at visible wavelengths using a plasmonic metasurface. *Light Sci. Appl.*, 3(5):e167, 2014.

[24]   E. Karimi, G. Zito, B. Piccirillo, L. Marrucci, and E. Santamato. Hypergeometric-Gaussian modes. *Opt. Lett.*, 32(21):3053–3055, 2007.

[25] S. N. Khonina, A. V. Ustinov, S. A. Fomchenkov, and A. P. Porfirev. Formation of hybrid higher-order cylindrical vector beams using binary multi-sector phase plates. *Sci. Rep.*, 8(1):1–11, 2018.

[26] M. Lax, W. H. Louisell, and W. B. McKnight. From Maxwell to paraxial wave optics. *Phys. Rev. A*, 11(4):1365, 1975.

[27] G. Machavariani, Y. Lumer, I. Moshe, A. Meir, and S. Jackel. Spatially-variable retardation plate for efficient generation of radially-and azimuthally-polarized beams. *Opt. Commun.*, 281(4):732–738, 2008.

[28] L. Marrucci, C. Manzo, and D. Paparo. Optical spin-to-orbital angular momentum conversion in inhomogeneous anisotropic media. *Phys. Rev. Lett.*, 96(16):163905, 2006.

[29] C. Maurer, A. Jesacher, S. Fürhapter, S. Bernet, and M. Ritsch-Marte. Tailoring of arbitrary optical vector beams. *New J. Phys.*, 9(3):78, 2007.

[30] M. A. A. Neil, F. Massoumian, R. Juškaitis, and T. Wilson. Method for the generation of arbitrary complex vector wave fronts. *Opt. Lett.*, 27(21):1929–1931, 2002.

[31] M. Neugebauer, S. Grosche, S. Rothau, G. Leuchs, and P. Banzer. Lateral spin transport in paraxial beams of light. *Opt. Lett.*, 41(15):3499–3502, 2016.

[32] J. F. Nye. Line singularities in wave fields. *Philos. Trans. R. Soc. Lond. A, Math. Phys. Eng. Sci.*, 355(1731):2065–2069, 1997.

[33] M. Padgett, J. Courtial, and L. Allen. Light's orbital angular momentum. *Phys. Today*, 57(5):35–40, 2004.

[34] S. Quabis, R. Dorn, and G. Leuchs. Generation of a radially polarized doughnut mode of high quality. *Appl. Phys. B*, 81(5):597–600, 2005.

[35] O. Quevedo-Teruel, H. Chen, A. Díaz-Rubio, G. Gok, A. Grbic, G. Minatti, E. Martini, S. Maci, G. V. Eleftheriades, M. Chen, et al.Roadmap on metasurfaces. *J. Opt.*, 21(7):073002, 2019.

[36] H. Rubinsztein-Dunlop, A. Forbes, M. V. Berry, M. R. Dennis, D. L. Andrews, M. Mansuripur, C. Denz, C. Alpmann, P. Banzer, T. Bauer, et al.Roadmap on structured light. *J. Opt.*, 19(1):013001, 2016.

[37] M. Sondermann and G. Leuchs. Photon-atom coupling with parabolic mirrors. In *Engineering the atom–photon interaction*, pages 75–98. Springer, 2015.

[38] M. S. Soskin, V. Denisenko, and I. Freund. Optical polarization singularities and elliptic stationary points. *Opt. Lett.*, 28(16):1475–1477, 2003.

[39] M. S. Soskin, V. N. Gorshkov, M. V. Vasnetsov, J. T. Malos, and N. R. Heckenberg. Topological charge and angular momentum of light beams carrying optical vortices. *Phys. Rev. A*, 56(5):4064, 1997.

[40] M. Stalder and M. Schadt. Linearly polarized light with axial symmetry generated by liquid-crystal polarization converters. *Opt. Lett.*, 21(23):1948–1950, 1996.

[41] P. Woźniak, P. Banzer, and G. Leuchs. Selective switching of individual multipole resonances in single dielectric nanoparticles. *Laser Photonics Rev.*, 9(2):231–240, 2015.

[42] N. Yu and F. Capasso. Flat optics with designer metasurfaces. *Nat. Mater.*, 13(2):139–150, 2014.

[43] N. Yu, P. Genevet, F. Aieta, M. A. Kats, R. Blanchard, G. Aoust, J.-P. Tetienne, Z. Gaburro, and F. Capasso. Flat optics: controlling wavefronts with optical antenna metasurfaces. *IEEE J. Sel. Top. Quantum Electron.*, 19(3):4700423, 2013.

[44] R. Zambrini and S. M. Barnett. Angular momentum of multimode and polarization patterns. *Opt. Express*, 15(23):15214–15227, 2007.

[45] Q. Zhan. Cylindrical vector beams: from mathematical concepts to applications. *Adv. Opt. Photonics*, 1(1):1–57, 2009.

8 Polarization of light at the nanoscale

In the last chapter we discussed that the polarization of a light beam can be structured and therefore also engineered spatially, paving the way for a large variety of applications. In this chapter we now go a step further and show that beside transverse components of the electromagnetic field, also field components aligned with the propagation direction may appear or, more appropriately speaking, are ubiquitous. The approximations discussed so far made us *blind* for those components, although for compliance with Maxwell's equations they are required. We now introduce how they come about, how they can be enhanced to even dominate electromagnetic field distributions, and which intriguing phenomena are linked to their appearance.

8.1 The ubiquity of longitudinal field components

If we revert briefly to Fig. 7.1 shown in the previous chapter we can see that, for the paraxial case, by definition, the beam converges or diverges only very slowly upon propagation. Nonetheless, the phase fronts are curved and the beam can also be represented by an angular bundle of rays all crossing at the geometrical focus point. Two of the limiting rays the beam envelope is asymptotically approaching when propagating towards $z \to \pm\infty$ are shown in the figure. Locally, the electric field vectors must be orthogonal to each of those partial rays, which causes a tilt of the electric field vector at this position of interest and with respect to the global coordinate frame. This is indicated in Fig. 7.1 (left) at a z-position far away from the waist plane. Hence, the electric field vectors (for rays which are not parallel to the optical axis) inherently also feature a non-zero component of the electric field oriented along the optical axis, which we will refer to as *longitudinal* field components. Similar arguments apply for the magnetic field in the corresponding plane of observation. This geometrical perspective already shows that the converging and diverging (diffracting) nature of light beams results in the appearance of longitudinal field components. We also see right away that the emergence of such components parallel to the optical axis of the light beam (mean propagation direction) is also dependent on the polarization state with respect to the propagation plane. If the beam in Fig. 7.1 was polarized orthogonal to the plotted plane of propagation, the electric field vector would not appear tilted in this very plane for geometrical reason. We will see later on in this chapter that indeed the polarization distribution of an input beam plays a major role in the process of focal field engineering (see Section 8.2.1).

The presence of longitudinal field components seems to be compliant with the previous statement that the solutions to the paraxial wave equation do not satisfy the full wave equation and Maxwell's equations. However, the geometrical picture used above does not equip us with a proper understanding of what to expect in the focal plane where the phase front is planar and the geometrical rays all come together. To

https://doi.org/10.1515/9783110668025-008

get a better understanding of effects to appear close to or in the focal plane, we assume a linearly (y) polarized light beam propagating along the z-axis in free space and plug it into Maxwell's equation(s). In preparation for this step, we rewrite Gauss' law in differential form for the vacuum [30]:

$$\nabla \cdot \mathbf{E} = 0 \quad \text{with } \mathbf{E} = (0, E_y, E_z)$$

$$\Rightarrow \quad E_z = -\int \left(\frac{\partial}{\partial y} E_y \right) dz. \tag{8.1}$$

Here we still assume for convenience that the x-component (crossed in-plane component) of the electric field is strictly zero. However, it should be noted here already that this is not the case for symmetry reasons (see Section 8.2.1). The equation above should help us now to characterize the longitudinal field component and its distribution in the focal plane ($z = 0$). For a beam exhibiting a fundamental Gaussian intensity envelope in the transverse plane,

$$E_y \propto e^{\frac{-(x^2 + y^2)}{w_0^2}} e^{ikz}, \tag{8.2}$$

to fulfill the aforementioned condition of a divergence-free electric field in vacuum, the transverse distribution of the longitudinal field components needs to take the following form in the focal plane:

$$E_z(x, y, 0) = -i \frac{2y}{k w_0^2} E_y(x, y, 0). \tag{8.3}$$

With $k = \frac{2\pi}{\lambda}$ we obtain

$$E_z(x, y, 0) = -i \frac{\lambda}{\pi w_0^2} y E_y(x, y, 0). \tag{8.4}$$

We can immediately identify various interesting features. First and foremost, the longitudinal field component is indeed non-zero (for $y \neq 0$). Furthermore, E_z features a Hermite–Gaussian transverse distribution depending on the distribution of the (main) y-component, which is parallel to the polarization. The strength of the longitudinal field component scales with the wavelength and inversely with the square of the beam waist w_0; the smaller the beam in the focal plane, the stronger the z-component. Last but not least, we also find that the transverse field component E_y and the longitudinal one E_z are $\pm \frac{\pi}{2}$ out-of-phase, as indicated by the imaginary unit i [6, 33]. The latter apparent detail holds immense potential in terms of fundamental physics and applications (see [1]) and will be discussed in more detail in Section 8.4.1.

In summary, we showed that based on this short theoretical calculation longitudinal field components must be present for the considered case (and many others) although their contribution might not be significant for large beam waists (limit of the paraxial approximation). However, they might grow to a substantial level quite rapidly in the case of tight focusing, discussed in the next section. We mention the work by Lax et al. who use a slightly different approach to also show that longitudinal electric fields must appear [23].

8.2 3D-Structured landscapes of light resulting from strong spatial confinement

As we have seen, longitudinal field components naturally arise from spatial confinement of electromagnetic fields. Light can be confined by different means, e. g., by focusing with a lens, by convergence, by transmission through tiny apertures or slits, and by interference. Furthermore, spatial confinement also naturally and inherently occurs for evanescent fields near interfaces. Evanescent fields resulting from total internal reflection at dielectric interfaces or surface plasmon polaritons propagating along metal–dielectric interfaces are just two examples. The evanescent decay with increasing distance from the surface they are bound to spatially confines the corresponding fields. In this section, we discuss two different and instructive scenarios, i. e., tight focusing in free space and evanescent fields, providing a basic framework and elaborating on some core aspects. The ultimate goal of this section is to emphasize how complex the structure of electromagnetic fields may become even under conventional circumstances and for relatively simple beams of light.

8.2.1 Tight focusing

In Chapter 7, we introduced an analytical framework for paraxially propagating light beams of different shapes and properties. The assumption that the light does neither converge nor diverge too quickly enabled us to retrieve solutions, well describing real light beams. However, this framework is neither appropriate nor accurate if we go beyond the paraxial limit, e. g., by focusing light tightly with a lens. Based on the discussion above (see Section 8.1), depolarization effects are expected to kick in rather drastically for strong confinement, which became apparent already when we discussed Eq. (8.4). To describe more accurately what effects shall be observed for tightly focused light, we need to apply a more capable framework not requiring any inadequate assumptions with respect to the beam propagation.

To date, various methods for calculating the focal field distributions in the limit of tight focusing have been introduced and applied. They provide an indispensable theoretical foundation for the quantitative utilization of tightly focused beams of light in countless applications in the fields of imaging, nano-metrology, advanced spectroscopy, and more. For instance, Lax et al. showed in 1975 that the assumptions made as part of the paraxial approximation are not fully consistent because it is assumed that the field is plane polarized while featuring a limited spatial extent [23]. They proposed extensions, or corrections, to the paraxial solutions for coping with this apparent issue. This ansatz provides for a more accurate description of the field.

Nonetheless, we want to focus here on an intuitive and powerful description, which is capable of also describing tight focusing by a lens. At its core, it is based on the viewpoint of geometrical optics. We know of course that geometrical optics in its original sense does neither provide information about the vectorial properties of a focal spot nor does it reflect properly how the beam propagates within or close to the focal volume. Furthermore, it even suggests an unphysical infinitely small focal area, or a focal point. The power of geometrical optics should, however, not be underestimated, like the case of ray-tracing shows impressively. It can serve as an underlying framework, which can be extended by involving polarization of light. For the discussion, we assume the following setting, also sketched in Fig. 8.1. A bundle of aligned parallel rays represented by an incoming wave vector $\vec{k}_{in}$ impinge on a thin lens (blue vertical line). This bundle of parallel rays can also be replaced by a single planar wave with wave vector $\vec{k}_{in}$ propagating along the z-axis. The incoming rays are refracted by the lens and redirected to meet at the geometrical focus point. Therefore, the field behind the lens cannot be described by a single planar wave anymore. Each ray now has its own wave vector $\vec{k}_{out,n}$ pointing towards the focal point with the limiting angle depending on the focusing strength and geometrical aperture size of the lens. Wave vectors above and below the optical axis (positive and negative y-axis) at equal distances feature the same z-component but opposite signs of the y-component. The plane waves propagating along each ray interfere and form the respective focal pattern in the focal volume similar to the interference pattern behind a double-slit or grating. The solid cone of rays and wave vectors is usually referred to as *angular spectrum*. The electric and magnetic field vectors for each geometrical ray or the corresponding plane wave have to be orthogonal to the ray or local wave vector due to the transversality condition. If the input field (planar wave) was vertically linearly po-

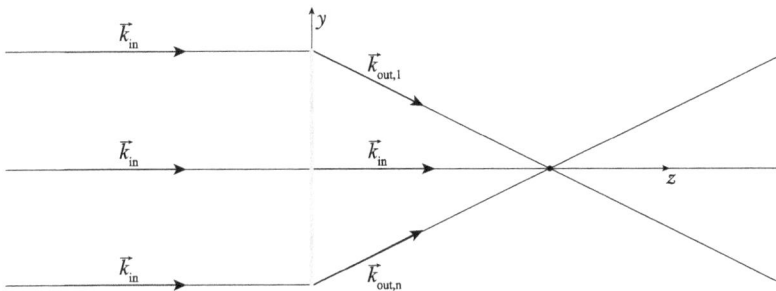

Figure 8.1: An incoming plane wave (or bundle of aligned parallel plane waves) impinges on a thin lens (blue vertical line), focusing individual light rays to meet at the geometrical focus. Each ray is redirected by the refraction in the lens. Hence, the propagation behind the lens cannot be represented anymore by a single $\vec{k}_{in}$ but an angular spectrum of wave vectors $\vec{k}_{out,n}$.

larized, the output polarization of each ray behind the lens will be dependent on the respective lateral position (along the x- and y-axis). This fact illustrates already that the lens (or non-paraxial propagation) induces a depolarization effect by redirecting the partial rays, which can also take a more complex form. A plane wave is an exact solution to the full wave equation and Maxwell's equations. Hence, also a superposition or spectrum of plane waves is a solution. We are consequently not applying any approximations till this point. We also see that due to the depolarization, the focal spot of course will not be infinitely small because the interference of plane waves will result in an interference pattern consisting of three-dimensional fields (electric or magnetic or both).

In the discussion we assumed that the lens is arbitrarily thin while still warranting a lens-like generation of a converging field (see Fig. 8.1). This would, however, require the rays originating from the rear surface of the lens to take different path lengths to reach the geometrical focus point. Different path length result in different phases across the aperture, influencing the focal interference. This is of course an artificial problem resulting from the chosen configuration. In a realistic lens, the light rays have to propagate through the lens material (glass or similar) with the thickness varying across the aperture. The optical path length thus depends on the position where the input ray hits the lens. Microscope objectives and lenses are usually designed such that they convert a planar incoming wave into a spherical wave converging towards the focus, warranting equal path lengths and no parasitic phase delays. This type of focusing element is usually called *aplanatic lens*. It fulfills *Abbe's sine condition* for a collimated input field:

$$\sin \theta = \frac{d}{f},\qquad(8.5)$$

with d the distance of the input ray from the optical axis, θ the corresponding angle of the resulting ray with respect to the optical axis, and f the focal length on the side of the image plane. An important aspect for the theoretical treatment of this conversion is also energy conservation, which must be taken into account.

Without going any deeper into these important but rather technical details, we now turn to the discussion of a powerful and versatile method, which is based on the aforementioned fundamental aspects for calculating focal field distributions. This method was introduced by Richards and Wolf in 1959 [33], extending the scalar diffraction theory. Their seminal work, which marks the hour of birth of vectorial diffraction theory, is still used heavily nowadays and can be implemented in a rather straightforward manner. The starting point for the derivation (not to be elaborated on here) is the angular spectrum introduced above. The electric field in real space can be represented by the field in momentum space (and vice versa) by the following integral assigning different wave vector components to the spectrum of plane waves (shown for propagation in z-direction):

$$\vec{E}(x, y, z) = \iint\limits_{-\infty}^{\infty} \vec{\tilde{E}}(k_x, k_y; 0)e^{-i[k_x x + k_y y \pm k_z z]} dk_x dk_y.$$

(8.6)

Richards's and Wolf's considerations result in the formulation of diffraction integrals [33], defining each field component analytically. In general, they can only be solved numerically. Two examples of focal fields resulting from tight focusing with high numerical aperture ($NA = 0.9$) lenses calculated using this technique are shown in Fig. 8.2. The numerical aperture NA of a focusing system is defined by $NA = n \cdot \sin\theta_{max}$, with n the refractive index of the medium behind the lens and θ_{max} the maximum focusing angle measured between steepest ray and optical axis. The results for a tightly focused y-polarized Gaussian input beam are presented in Fig. 8.2 (a). The input intensity and polarization defines the corresponding amplitude and polarization for each ray impinging on the lens in the geometrical picture discussed above. All presented focal components are normalized with respect to the maximum value of the total electric energy density distribution (left). As we can see clearly, several effects can be observed, which are all a direct consequence of the strong spatial confinement. The total electric energy density is elongated along the axis of the input polarization (here, the y-axis). The main contributors to this elongation are the transverse y-component and the longitudinal z-component of the field. The latter, for instance, features a two-lobe pattern along the original polarization axis. As a result of previous discussions, the appearance of such longitudinal field components and their pattern are not surprising anymore. However, we see that the focal fields follow more sophisticated distributions than approximately derived above because they are not (Hermite–) Gaussian anymore (see Eq. (8.4)). All components are actually surrounded by weaker side-lobes. The focal field also features a non-zero in-plane (x) component. It is relatively weak in comparison to the other components and therefore not visible for the chosen normalization. The phase distribution shown as an inset reveals the structure of this component. The four main lobes show a clover-leaf-like distribution.

For comparison, the results for a radially polarized input beam are shown in Fig. 8.2 (b). For radial polarization, the input electric field is locally pointing in the radial direction. Here the doughnut-shaped input intensity profile (similar to the case of azimuthal polarization in Fig. 7.3 (a)) is taken into account together with the position-dependent polarization direction in the angular spectrum representation above. As we can see, the total electric energy density distribution (left) as well as all individual components take a fundamentally different shape in comparison to a Gaussian input beam. As a direct consequence of the cylindrical symmetry of the input beam, also the total electric energy density features a circular shape. It also exhibits a distribution peaking on the optical axis whereas the input beam was hollow. All electric field components are rotated and tilted forward by the lens. When interfering in the focal

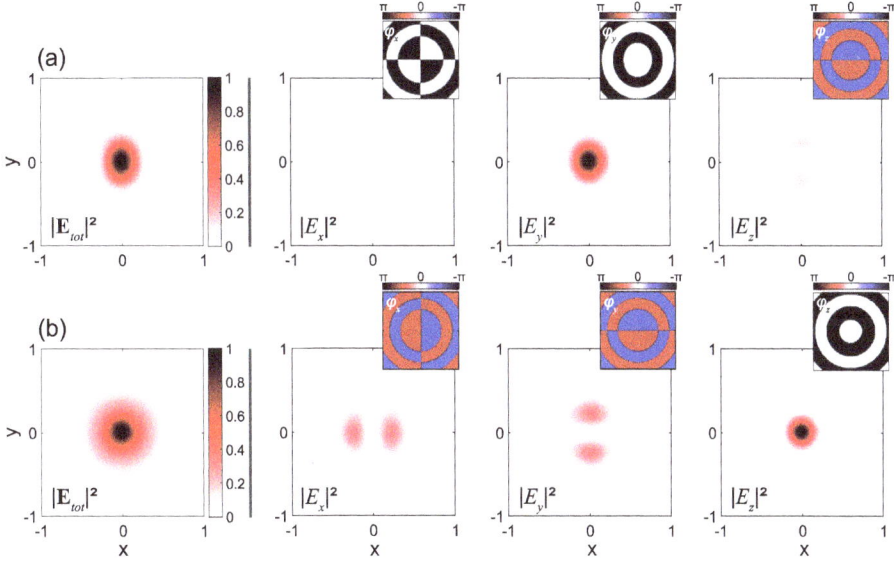

Figure 8.2: Left: Total electric energy density distribution in the focal plane for tight focusing. Center and right: energy density distributions of all electric field components. Maps of relative phases are shown as insets covering the same axes dimensions. Two cases are shown: tightly focused (a) linearly polarized Gaussian beam and (b) radially polarized doughnut mode. All distributions are normalized to the maximum of the total electric energy density. Reproduced from Ref. [4].

volume, the transverse electric field components on the optical axis cancel for symmetry reasons while the longitudinal components interfere constructively to form a strong on-axis component (see also Fig. 8.2 (b), right). The central z-component is surrounded by ring-shaped transverse field components. In the focal plane we therefore find a three-dimensional electric field with the polarization depending on the actual position with respect to the optical axis. Based on these two examples it becomes clear that the focal field can be engineered by modifying the input intensity, polarization and phase distributions.

Before we move on to discuss in more detail other configurations and related phenomena, we want to allude to an intriguing aspect with respect to the *democracy* of electric and magnetic fields. The energy density distributions of electric and magnetic fields (or electric and magnetic field intensities) for plane waves or paraxial beams of light are equivalent. This is a direct consequence of the fact that their electric and magnetic fields are locally orthogonal to each other. For instance, while the electric field of a radially polarized light beam in the paraxial domain is oriented like the spokes of a wheel, the magnetic field is pointing locally in a direction tangential to the *tire* or along the azimuthal angle. Both of them equal zero on the optical axis and therefore form doughnut-shaped field intensity distributions. However, the situation is drastically different for light fields described in a full (Maxwellian) and complete framework without the assumption of paraxial propagation in place. When comparing electric with magnetic field distributions, for instance in the focal volume of a lens, the differences in the field intensity maps are striking. The electric field for the radially polarized beam lie in the propagation plane spanned by the input ray impinging on the lens and the ray leaving the lens redirected to propagate towards the geometrical focus. This configuration is sometimes also referred to as *transverse magnetic (TM)* polarization. Hence, the magnetic field is locally orthogonal to the electric field vector and the plane described above. As a direct geometrical consequence, the electric field is rotated to remain orthogonal to the local wave vector, as discussed before, while the magnetic field component does not experience any rotation. The magnetic field is hence purely transverse in the focal plane while the electric field features transverse and longitudinal components different from zero. Although this phenomenon follows from simple geometrical considerations, it is undoubtedly quite astonishing that the electromagnetic field can be purely electric (or magnetic) for certain points (or lines like the optical axis) in space for a rather conventional setting, i. e., free-space propagation. Furthermore, the distributions of electric and magnetic fields can be substantially different, a feature which is very much common for other settings such as standing waves or evanescent fields appearing near interfaces (see the next sections). The input polarization-selective appearance of longitudinal fields has also far-reaching effects with respect to other field parameters, such as the spin density, as discussed below or in Refs. [2, 29].

8.2.2 Near fields and evanescent waves

The discussion of light's structure can also be extended to electromagnetic fields near interfaces, scatterers, diffractive elements or similar inhomogeneities light is interacting with in space. The electromagnetic field close to refractive index steps, metal interfaces or similar naturally features an evanescent contribution. From a more general perspective, evanescent fields need to be considered whenever space is inhomogeneous in a certain sense. Evanescent fields decay exponentially in amplitude in contrast to propagating waves. To illustrate the importance of evanescent fields, their inherently structured nature, and their strong contribution to the near field, it is instructive to consider the simple and very fundamental example of a point-like electric dipole oscillating at a given frequency ω_0. When observed from a sufficiently large distance ($\gg \lambda_0 = 2\pi c/\omega_0$) and at an angle of 90° with respect to the orientation of the dipole moment, the emission can be considered to be plane-wave-like. Furthermore, when resolved angularly, the emission pattern features the well-known sine-squared intensity distribution with zero emission along the dipole moment (dipole axis). However, the situation is considerably different in close vicinity to the dipole. A charge oscillating along the dipole axis creates a highly structured near-field distribution resulting from the strongly curved field lines. The evanescent field contributions originating from the charge(s) die out quickly with increasing distance. However, and in contrast to the far-field distribution discussed above, which features no intensity along the dipole axis, the near field peaks at those positions. This effect can be attributed solely to the evanescent fields. This simple example visualizes the substantial differences between near- and far-field intensity distributions and the highly structured nature of evanescent waves. Now, we move on to the discussion of the polarization of these near fields. Caused by its strong spatial dependence in close vicinity to the dipole, the polarization also varies substantially on a small scale (smaller than the wavelength corresponding to the oscillation frequency of the dipole). When the charges are separated, Maxwell's equations tell us that we should expect a gradient of the electric field surrounding the charges. While the charges flow along the dipole axis (just like in an antenna), a magnetic field arises, curling around the current. The resulting fields oscillate in time with the dipole frequency. The time-oscillating magnetic fields result in a curl of the electric field and so forth. The polarization therefore varies on a very small length scale.

Another important case to be shortly discussed here is the evanescent field at dielectric interfaces, for instance between air and a dielectric material such as glass. If a plane wave impinges from the glass side of the interface at an angle greater than the angle of total internal reflection, it will be fully reflected. No propagation component of the field will be found in the air half-space. However, the field is also not strictly zero there. As a direct consequence of the boundary conditions, the field does not drop to zero abruptly, but decays exponentially with increasing distance from the interface. The evanescent wave propagates along the interface and it is spatially confined to it. The electric field is parallel to the propagation plane (spanned by the input and output ray) while the magnetic field is orthogonal to it. As we will see below in Section 8.4.1, the electric field of evanescent waves features an intriguing property with respect to the polarization.

8.3 Measuring structured light at the nanoscale

With the complexity, three-dimensional nature, and sub-wavelength features of highly confined electromagnetic fields, conventional measurement techniques for beam inspection and the measurement of polarization as well as intensity distributions are neither appropriate nor applicable anymore. Conventional measurement tools, such as cameras, lack in sufficient spatial resolution to resolve the substructure of the field, in sensitivity to distinguish between transverse and longitudinal field components, and the capability to measure the relative phase between components simultaneously. With the increasing importance of nanoscale structured light in cutting-edge research and modern applications, also the need for versatile and powerful characterization methods for complex fields is growing.

Multiple experimental methods have been introduced, tested and discussed in the literature, which allow for the measurement of electric field intensity distributions, for the study of individual field components, or even the reconstruction of the full-field information (amplitude and phase distributions of all field components) of propagating or evanescent light with deep sub-wavelength spatial resolution. Below, we discuss briefly some selected methods capable of providing quantitative access to nanostructured propagating or evanescent fields.

8.3.1 Probing spatially confined fields

In this Section, which has more a review-type rather than a didactic style, we want to give a brief overview with respect to experimental techniques for probing the electromagnetic field at nanoscale dimensions. As mentioned before, it seems to be intuitive to utilize probe-based scanning approaches for the measurement of confined beam profiles instead of using detectors with insufficient spatial resolution and missing phase and polarization sensitivities directly. In fact, most techniques presented to date feature probes locally interacting with the light field. The level of information gained by such measurements strongly depends on probe design and analysis strategy. To gain, for instance, information about the relative phase of the individual field components in a spatially resolved manner, usually additional measures have to be taken to provide a phase reference just like in an interferometer.

For the sake of convenience, we focus here on a small selection of powerful methods to be discussed in a bit more detail. However, in passing, we plan to mention briefly some other methods as well. We start with a discussion of conventional techniques for profiling relatively large beams.

Many beam profilers available commercially nowadays are based on simple high-resolution cameras, which allow for measuring the intensity profile of light beams. With ever decreasing pixel sizes and pitches, the intensity distribution can be measured accurately, even for small beam diameters on the order of tens of micrometers. If we are interested in the polarization distribution of paraxial beams of light, such cameras can be combined with polarizing elements (see Chapter 9) to perform a spatially resolved Stokes measurement, which defines the polarization distribution in the beam cross-section in addition to the intensity profile. Furthermore, polarization cameras have been introduced, which feature linear polarizers in front of the pixels to distinguish between different linear polarization states. If on top of the polarization also the phase front is of interest, the complexity of the measurement procedure is increasing because additional measurement devices, such as phase front sensors, are required or interferometric schemes have to be implemented. Phase, polarization and intensity distributions fully characterize the beam.

As an alternative to a camera-based measurement of the beam profile, also photodetectors combined with moving mechanical elements (see Fig. 8.3), e. g., slits or pin-

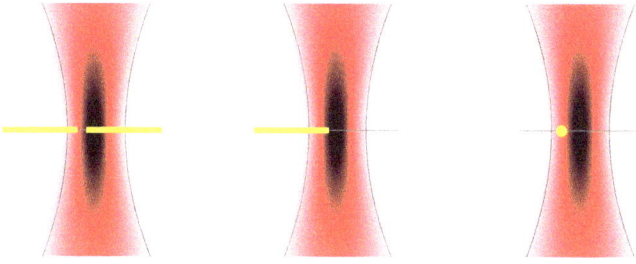

Figure 8.3: Different types of probes for profiling focused beams of light. From left to right: pinhole or slit, knife-edge, and (nano-)particle.

holes in opaque films, sharp metal edges (*knife-edges*) or fluorescent point-like probes [15, 35] are used and have been partially commercialized as beam profilers. The transmitted power is measured with a photo-detector and the profile can be reconstructed tomographically from the position-dependent scan data. Such devices usually only measure the intensity profile or certain beam parameters like the radius. The reader might wonder now, whether or not such schemes can also be applied if non-paraxially propagating or tightly focused light beams are to be profiled. In the previous sections it became clear that beside a structure to the transverse electromagnetic field, strongly confined light fields will feature complex three-dimensional distributions with spatially varying phase, polarization and intensity. It is therefore immediately apparent that the aforementioned methods for beam profiling might not be sufficiently capable or have to be adapted. An instructive example in this context is indeed the knife-edge-based method. It utilizes an opaque layer with a sharp edge (see Fig. 8.3, center), which is moved across the beam to be profiled, to block the beam partially. From the photo-current curve recorded by a photo-detector, the beam profile projection along the scan direction can be reconstructed. To adapt this scheme to also work for tightly focused light beams, a number of modifications are necessary to reduce the amount of artifacts and errors. Because a tightly focused light beam diverges quickly after trespassing the focal plane, the detector should be placed very close to the knife edge itself. In addition, the steep wave vectors involved in tight focusing require an ultra-thin footprint of the sharp-edged material layer forming the knife edge, while still being opaque. Experimentally, this was realized by fabricating knife edges directly on top of detectors and based on thin metal films. However, additional problems arise from the complex interaction of focused beams with metal edges. For instance, the sharp knife edge will interact with the impinging light in a polarization sensitive manner, e. g., differently for light polarized along the edge or orthogonal to it. This polarization dependence can influence the resulting beam profile in a parasitic way, deforming, shifting or skewing it [20, 27]. The impact on the measured data depends on various parameters, for instance on the edge material [14, 27, 32], the substrate or the detector material underneath [21]. The artifacts introduced by the measurement can be compensated for by a proper data post-processing strategy [20] or circumvented by an appropriate choice of the involved materials [14, 32]. The knife-edge method therefore turns into a powerful tool for reconstructing the total electric field intensity profile of even tightly focused light beams. But, so far, its capabilities are limited to intensity distributions.

To get access to individual field components and their energy density distributions, the aforementioned method of fluorescent point probes has been adopted and modified accordingly. A tightly focused light beam is scanned across randomly oriented dye molecules embedded into a transparent dielectric matrix [31, 35, 37]. Although introduced as a method for molecule orientation sensing, the patterns observed in a fluorescence confocal scanning microscope also allow for retrieval of the focal field distributions (amplitude distributions of field components) if various molecules of different orientation are scanned.

An alternative version of this method can also enable the measurement of relative phase maps. For this purpose, various groups have utilized near-field scanning optical microscopes (NSOM) [38]. A detailed discussion of NSOMs would go beyond the scope of this book. In short, NSOMs are based on sharp metal tips brought into close vicinity of scatterers, waveguides or interfaces to pick up the near-field information (by scattering or propagating it to the far field). NSOMs are excellently suited for measuring the complex near-field polarization distribution and other field parameters [34]. Furthermore, they also have been successfully applied for measuring focal fields [19, 25]. For granting access to polarization and phase information, the implementation of additional polarization elements and a phase reference is required.

We complete our overview by discussing a more recent technique for the full reconstruction of focal fields, not requiring the implementation of any additional phase reference, polarization analysis, or similar. This method is based on a very intuitive approach and probe-design, while enabling the reconstruction of both amplitude and phase distributions of the individual electric field components. It utilizes a nanoparticle acting as a scanning probe (see Fig. 8.3, right). The nanoparticle supports multipolar resonances and is immobilized on a dielectric substrate. It is placed in the focal plane of a tightly focused light beam under study, and raster-scanned across the latter [5, 6]. It might sound surprising that such a seemingly simple scheme is sufficient to fully reconstruct the electric field in its amplitude and phase distributions. As elaborated on in the above-mentioned references, the key ingredients to this technique enabling also access to the phase are a rigorous theoretical backbone and the measurement of transmitted (or reflected) and scattered light in an angularly resolved fashion. No additional polarization or phase measurement apparatus is required to analyze the light after it has interacted with the sample. The intensity distribution is measured with a camera, which images the back focal plane of a collecting microscope objective. This distribution contains information about the scattered part of the light field, the input beam itself, and their interference. Hence, the interference term carries the information about the phase of the excitation field components. With the probe interacting locally with the excitation field under investigation, the desired information about the local field can be retrieved. From a theoretical perspective, the power detected in the back focal plane (resulting from an integration across different areas of the intensity distribution for each particle position) can be decomposed into input power, scattered power and the aforementioned interference term. Scattered and input fields are related to each other via a scattering matrix (T-matrix), which contains information about the scatterer and the substrate underneath. The fields themselves can be represented by a series of vector-spherical harmonics, which represent multipoles (dipoles, quadrupoles etc.). The latter is a natural basis for this kind of interaction because it reflects the optical response of the probing particle. It can be shown that a limited number of multipole orders are sufficient together with an adapted number of particle steps and step-sizes. Furthermore, the integration limits in the measured back focal plane images (angular spectrum) have to be chosen accordingly, representing certain solid angles within which the detected power is taken into account for the analysis. As a result, the input field in the probed plane can be retrieved accurately with deep sub-wavelength spatial resolution. The achievable resolution depends on the aforementioned parameters. With the experimental reconstruction of amplitudes and phases, the complex field structure at nanoscale dimensions becomes experimentally accessible [5, 6]. This method can be used to characterize tightly focused light beams used as tools for nano-optics experiments, for analyzing focusing systems, and many more.

In summary, the last decade has seen a promising development of novel techniques and refined methods with proven capabilities. They allow for accessing the sub-wavelength features and the three-dimensional nature of highly confined electromagnetic fields and their complex polarization distributions.

8.3.2 Extended Stokes parameters for three-dimensional fields

On the theoretical side, also the previously introduced frameworks for the description of polarization and its spatial distribution, such as Stokes parameters, angular momenta, and topological features, need to be revisited. The latter aspects will be covered below while we focus first on Stokes parameters.

By definition, the Stokes parameters as introduced in Section 3 are based on the assumption that the electric field is restricted to a plane orthogonal to the propagation direction. More generally speaking, they can be applied if the field is polarized in a plane. This assumption is perfectly valid for plane waves or paraxially propagating beams of light. However, the electromagnetic field is usually inherently three-dimensional, while especially in the case of strong spatial confinement (tight focusing, evanescent fields, etc.), longitudinal components contribute substantially to the total field. The Stokes formalism thus has to be extended to cover the full extent of the field. This can be done in a very intuitive and convenient manner also for fully polarized three-dimensional fields, following the original idea behind the Stokes vector and its components. To cover the full extent of the three-dimensional field, the three orthogonal (electric) field components E_x, E_y, and E_z need to be compared to each other with respect to their field intensities and phases, just like before in the case of conventional Stokes parameters and purely transverse fields. It is convenient to start again with the Jones vector, while we follow the discussion and notation in References [12, 36]. The Jones vector for a fully polarized three-dimensional field now takes the form of a three-components column vector and reads as follows:

$$\vec{E} \equiv \begin{pmatrix} E_x \\ E_y \\ E_z \end{pmatrix} = \begin{pmatrix} E_{0x}e^{i\delta_x} \\ E_{0y}e^{i\delta_y} \\ E_{0z}e^{i\delta_z} \end{pmatrix}. \tag{8.7}$$

The phases can be defined with respect to an absolute phase reference, reducing the number of independent variables to five (including a total intensity reference). Although the field is now allowed to oscillate in three-dimensional space with no physical restriction to the transverse plane with respect to the propagation direction, the electric field can still be described, in the most general case and for fully polarized light, by a polarization ellipse (or line/circle for the limiting cases) just like in the two-dimensional case discussed in Section 3.3. Starting with the generalized Jones vector, we can now also write down the generalized Stokes parameters Λ_i for a fully polarized field as follows:

$$\begin{aligned}
\Lambda_0 &= E_x^2 + E_y^2 + E_z^2, \\
\Lambda_1 &= 3E_{0x}E_{0y}\cos(\delta_y - \delta_x), \\
\Lambda_2 &= 3E_{0x}E_{0y}\sin(\delta_y - \delta_x), \\
\Lambda_3 &= \frac{3}{2}(E_x^2 - E_y^2), \\
\Lambda_4 &= 3E_{0x}E_{0z}\cos(\delta_z - \delta_x), \\
\Lambda_5 &= 3E_{0x}E_{0z}\sin(\delta_z - \delta_x), \\
\Lambda_6 &= 3E_{0y}E_{0z}\cos(\delta_z - \delta_y), \\
\Lambda_7 &= 3E_{0y}E_{0z}\sin(\delta_z - \delta_y), \\
\Lambda_8 &= \frac{\sqrt{3}}{2}(E_x^2 + E_y^2 - 2E_z^2),
\end{aligned} \tag{8.8}$$

with

$$\Lambda_0^2 = \frac{1}{3}\sum_{i=1}^{8}\Lambda_i^2. \tag{8.9}$$

We see that we have nine generalized Stokes parameters together with a relationship connecting all parameters comparable to that for the two-dimensional case. However, we noted above that the three-dimensional field can be defined by five independent parameters in the Jones vector. It can be shown that the generalized Stokes parameters are not all independent of each other and that a set of five is sufficient to describe the field's behavior. For more details and possible combinations of Stokes parameters allowing for the full characterization of the field, the interested reader should refer to Ref. [36]. The parameters listed above can be used to represent the local polarization in fully polarized three-dimensional fields. However, their measurement is considerably more complicated than the measurement of the conventional Stokes parameters in paraxial light (see discussion in Section 8.3).

In the following section we will see that the three-dimensional character of spatially confined fields evokes intriguing and fascinating phenomena connected to polarization in general as well as spin angular momentum and the topological structure of light in particular.

8.4 Exotic phenomena based on polarization effects in spatially confined light

The presence of longitudinal field components, either electric, magnetic or both, gives rise to interesting phenomena linked to the structure of the field. Based on the discussion above we noticed already that field intensity distributions may change substantially within small spatial scales if light is highly confined, in striking contrast to paraxially propagating beams. An instructive example was radially polarized light forming a ring of intensity in the paraxial limit, while featuring an on-axis maximum for tight confinement by a lens. However, while discussing the origins and manifestations of longitudinal fields, we also came across additional peculiarities with respect to the dynamics of the fields and their spatial features.

In this section, we now elaborate on some selected phenomena related to polarization and to be discovered in three-dimensional electromagnetic fields, i. e., the occurrence of transverse spin and topological properties.

8.4.1 Transverse spin

In the previous chapter we mentioned already that the polarization is also intimately connected to the spin angular momentum of light. The spin per photon reaches values of up to $\pm\hbar$ for circularly polarized light and is identical 0 for linear polarization. The spin angular momentum is an integrated quantity resulting from the spatial integration across the spin density distribution $\vec{s}$ of, e. g., a light beam. The spin density defines the local spin of the field and, following the notation shown in [1, 7], it is given by

$$\vec{s} = \vec{s}_E + \vec{s}_H = \Im\mathfrak{m}(\epsilon_0 \vec{E}^* \times \vec{E} + \mu_0 \vec{H}^* \times \vec{H})/4\omega. \tag{8.10}$$

Here $\vec{s}_E$ ($\vec{s}_H$) correspond to the spin density of the electric (magnetic) field. The spatial dependence of this quantity is omitted for the sake of brevity. We focus in our discussion on the electric part of the spin density. The spin density is strictly zero if the field is locally linearly polarized, while it takes non-zero values for elliptical polarization, and reaches its maximum (minimum) for circular polarization. The spin density is proportional to the parameter c as defined in Eq. (7.19). In the case of paraxial light or individual plane waves, the electric and magnetic fields are restricted to the transverse plane, which forces the spin density to be a scalar number in principle (only the s_z component can be different from zero). Hence, the spin is always aligned with the propagation axis (parallel or anti-parallel) for such cases; the spin is longitudinal. However, if we allow for three-dimensional electric and magnetic fields as they appear in focused light beams or other scenarios, also the spin density in Eq. (8.10), will be a vector with three entries. From its construction above it becomes apparent that also transverse components of the spin density (s_x^E and s_y^E) may appear as long as the corresponding field components, i. e., E_y and E_z as well as E_x and E_z, respectively, are non-zero and appropriately de-phased. In the discussion of tightly focused light beams (Fig. 8.2) and also in the introductory section, Section 8.1, we learned already that by spatially confined light beams naturally feature longitudinal together with transverse field components [6]. The only missing ingredient for the spin to density have non-zero transverse components, and for the field to be elliptically or circularly polarized in the propagation plane, is an appropriate phase relation (insets in Fig. 8.2). If we re-visit again the focal distributions shown in this figure, we can see that, for the shown cases also the remaining requirements for non-zero transverse spin components, i. e., a relative phase of $\pm\pi/2$, are fulfilled. In Fig. 8.4, we show the focal spin density distributions (transverse components only) for the cases of tightly focused linearly and radially polarized light beams, respectively [1, 28]. We can clearly see that the transverse components of the spin density are non-zero for certain areas where distributions of longitudinal and transverse field components overlap. In addition, also their sign is position-dependent. The de-phasing of the field components responsible for the appearance of a non-zero transverse spin density can be explained in a simple and intuitive manner by the Gouy phase discussed above in the context of paraxial light propagation. Also for tight focusing, the field accumulates a phase delay (with respect to a planar reference wave). The total phases accumulated while propagating towards the focal plane depend on the mode order or spatial profile of the field. The longitudinal field component of, e. g., a tightly focused radially polarized beam features for symmetry reasons a different spatial profile than the transverse field. Hence, the respective field components are de-phased differently, resulting in an effective $\pm\pi/2$ phase difference in the focal plane. This relative phase together with the spatially partially overlapping field distributions result in non-zero transverse spin density components with the corresponding elliptically or circularly polarized field spinning in the propagation plane (see Fig. 8.5).

It is also worth noting that the longitudinal component of the spin density s_z for the presented cases is strictly 0 everywhere, which can be understood by looking at

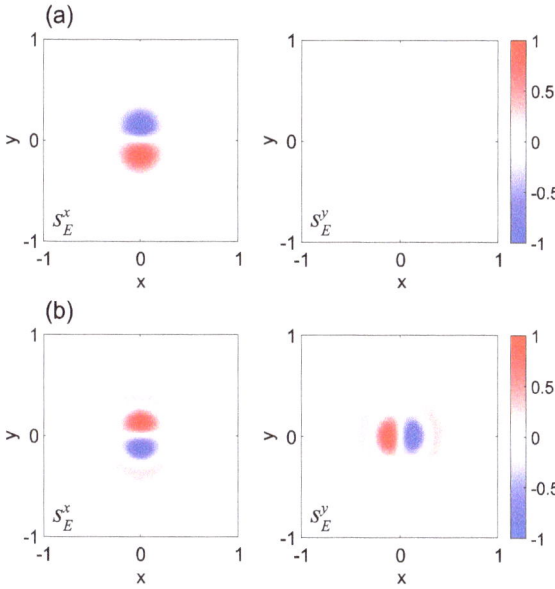

Figure 8.4: Transverse focal spin density components s_x^E and s_y^E for a tightly focused (a) linearly polarized Gaussian beam and (b) radially polarized doughnut mode. All distributions are anti-symmetric with respect to either the *x*- or *y*-axis. Reproduced from Ref. [4].

Fig. 8.2. We also note that for the chosen input beams the distributions are all anti-symmetric with respect to one of the coordinate axes spanning the transverse plane (x or y). Hence, the integrated quantity $\vec{S}_E$ at $z = 0$, defined by

$$\vec{S}_E = \iint\limits_{-\infty}^{\infty} \vec{s}_E \, dx \, dy,\tag{8.11}$$

also called the net spin, is strictly zero in all its components for the presented and many other cases. As a side note we mention here briefly that also beams carrying non-zero transverse components of the *net spin* can be constructed. They feature transverse spin density distributions, which do not integrate to zero and therefore are not anti-symmetric. For more details the interested reader is referred to Refs. [1] and [3].

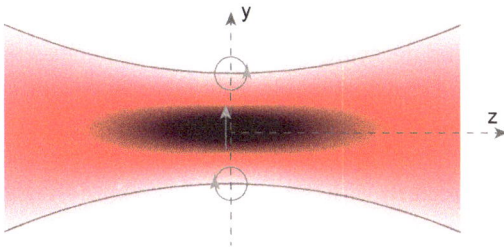

Figure 8.5: Illustration of the local focal electric field for a focused Gaussian light beam polarized along the *y*-axis. A snapshot of the electric field is shown for three positions in the focal plane (on the optical axis as well as above and below it). It is linearly polarized on the optical axis and circularly (or elliptically) polarized along the *y*-axis.

The reader might be of the impression now that transverse spin only appears in the tight focusing regime. However, if we recall the discussion in Section 8.1, longitudinal field components (electric or magnetic) are ubiquitous and appear also for weakly focused beams to be compliant with Maxwell's equations. However, for a collimated beam, the contributions of longitudinal fields is negligibly small for geometric reasons, and, thus, the transverse spin density components are small as well.

In general, the phenomenon of transverse spin densities is not exclusive to free-space focusing. It can also be found in either the electric or magnetic field or both for other types of spatial confinement. Interesting and important examples are evanescent waves at dielectric interfaces [9], or equivalently, propagating waves at metal–dielectric surfaces (propagating surface plasmon polaritons; SPPs) [10]. Both types of waves are confined to a volume close to the surface (exponential decay along the surface normal) while propagating along it. For a better understanding, we again turn back to the discussion of evanescent waves at dielectric interfaces (see also Section 8.3.1). Transverse spin density components are equivalent to an elliptical or circular polarization component with the field spinning in the propagation plane (plane spanned by transverse and longitudinal axis). To qualitatively predict the polarization of an evanescent wave resulting from total internal reflection at, e. g., a glass-air interface, we discuss the process of a plane wave totally reflected more carefully. For the discussion, the wave impinging above the critical angle is set to be in-plane linearly polarized. The wave interferes partially with the reflected copy of itself. The reflected planar wave has the same amplitude but it is slightly shifted in its phase as a direct consequence of the Fresnel coefficients. Both waves—incoming and reflected—are superposed and interfere partially (if their polarization states are not fully orthogonal, i. e., for an angle of incidence different from 45°). The phase delay introduced by reflection (Fresnel coefficients) is strongly angle dependent and changes from 0 to π between the critical angle and grazing incidence (90°). The resulting polarization of the two-plane-wave pattern in the glass half-space shows therefore a complex polarization distribution varying with increasing distance from the interface also defining the relative path difference. Close to the interface in the optically denser medium (glass) where the path difference goes to zero, the field is elliptically polarized in the plane of incidence, the signature of a non-zero transverse spin density (of the electric field). On the other side of the interface, in the air half-space, the field is also elliptically polarized as a consequence of the continuity conditions for the electric field (and displacement field) components at the interface. Hence, the evanescent wave traveling along the interface and decaying exponentially with increasing distance to it, is elliptically polarized in the propagation plane and therefore features also a transverse spin density different from zero.

In conclusion, the appearance of transverse spin, and equivalently, electric and magnetic fields polarized elliptically or circularly with the polarization ellipse lying in the meridional plane is a ubiquitous phenomenon, although introduced and put on a solid theoretical foundation related to angular momenta only very recently [1, 11].

Transverse spin is one of the key features enabling interesting applications in light-routing, nano-metrology, quantum optics and more [1, 26].

8.4.2 Topological features of confined light

In the previous chapters, the appearance of topological features in light fields was discussed already in the context of phase vortices and polarization singularities in the scalar case or for generic two-dimensional ellipse fields. By definition, the corresponding electromagnetic fields were two-dimensional evolving in the same plane. We now want to briefly extend this discussion to three-dimensional field distributions. We will see that these may host exotic and surprising topological structures. The phenomena that follow closely resemble the geometric-phase effects described in Chapter 6.

The aforementioned polarization distributions (ellipse fields) evolving around C-points form topological structures with a topological index defining their type. Naturally, the addition of an extra field component orthogonal to the plane lifts the field out of the plane and creates a three-dimensional landscape of the field in the given plane. Isaac Freund was one of the first to study corresponding systems [16, 17]. His studies showed that, for instance, the major axes of the polarization ellipses traced along closed lines around C-points can show intriguing topological structures. In particular cases, the major axis twists and turns around the chosen trace and when returning to the starting point, the number of turns will be equal to $m/2$ with m an odd integer. This behavior is similar to the case of two-dimensional ellipse fields (see Fig. 7.4) where the ellipse rotated in plane by $\pm180°$. However, with the rotation now happening in 3D space, the resulting structure formed by the major axes of the traced polarization ellipses features the shape and properties of a *Möbius* strip. A Möbius strip is an object, which possesses only one edge and one surface by construction (see Fig. 8.6). It is stable under deformations and can be classified as a topology. As shown in Fig. 8.6, it can be constructed easily from paper by twisting one end of a paper strip by 180 degrees and taping both ends together. By construction, the originally separated surfaces of the paper strip are now connected. The same is true for the two (long) edges.

In his work, Freund also proposed a method for creating such elusive strips of light. His suggestion was based on the interaction of two orthogonally polarized (left- and right-handed circular) non-coaxially propagating light beams of different spatial phase structures (different azimuthal phase ramps). In his case, a fundamental Gaussian beam and a Laguerre–Gaussian beam of first azimuthal order were chosen. When studied in the plane of interaction, the field would naturally feature also field components orthogonal to this plane. The differences in the azimuthal phase change and the opposite handedness of polarization give rise to the appearance of Möbius strips around a central C-point. In 2015, Bauer et al. proved the existence of polarization Möbius strips experimentally. In their case, they created the topological structures requiring three-dimensional field distributions in a different and more efficient manner.

Figure 8.6: A half-twist Möbius strip constructed from a colored sheet of paper. The two surfaces of the paper strip are indicated in orange and yellow, whereas the two long edges of the original strip are colored in dark-blue and green. By twisting one end of the paper strip by 180 degrees and gluing both ends together, the blue edge is connected to the green one and the orange surface to the yellow surface.

They chose the scheme of tight focusing of co-propagating light beams. Upon tight focusing, strong longitudinal field components appeared, which together with the chosen input states generated optical polarization Möbius strips formed around the optical axis (C-point) in the focal plane [5]. A particle-based probing technique [6] allowed them to measure the full field in the focal volume (see Section 8.3). The orientation of the major ellipse axes can be calculated using Eq. (7.19). Two years later, Galvez and coworkers also realized the scheme originally proposed by Freund and measured the strips for this configuration [18].

Another topological phenomenon, which we mention here briefly and which may be found in scalar or polarization varying field distributions, are so-called knots. Dennis, Berry, Padgett and others studied these intriguing and mind-boggling structures in theoretical and experimental detail [8, 13, 24]. Knots can be formed by phase singularities in three-dimensional space or also by the polarization (e. g. knotted C-lines). In the latter case, points of conventional circular polarization (with the field spinning in the transverse plane) form closed lines in a given volume of a propagating beam [22]. These lines turn out to be knotted under certain circumstances. The appearance of knotted polarization structures is again linked to different phases accumulated upon propagation of modes of different order. These additional examples showcase again the richness of electromagnetic fields.

Bibliography

[1]   A. Aiello, P. Banzer, M. Neugebauer, and G. Leuchs. From transverse angular momentum to photonic wheels. *Nat. Photonics*, 9(12):789–795, 2015.

[2]   A. Aiello and M. V. Berry. Note on the helicity decomposition of spin and orbital optical currents. *J. Opt.*, 17(6):062001, 2015.

[3]   P. Banzer, M. Neugebauer, A. Aiello, C. Marquardt, N. Lindlein, T. Bauer, and G. Leuchs. The photonic wheel—demonstration of a state of light with purely transverse angular momentum. *J. Eur. Opt. Soc., Rapid Publ.*, 8(0), 2013.

[4]   P. Banzer. Nano-optics and plasmonics with complex spatial modes of light—structured electromagnetic fields at the nanoscale, 2019. Habilitation Thesis.

[5]   T. Bauer, P. Banzer, E. Karimi, S. Orlov, A. Rubano, L. Marrucci, E. Santamato, R. W. Boyd, and G. Leuchs. Observation of optical polarization Möbius strips. *Science*, 347(6225):964–966, 2015.

[6]   T. Bauer, S. Orlov, U. Peschel, P. Banzer, and G. Leuchs. Nanointerferometric amplitude and phase reconstruction of tightly focused vector beams. *Nat. Photonics*, 8(1):23–27, 2014.

[7]   M. V. Berry. Optical currents. *J. Opt. A, Pure Appl. Opt.*, 11(9):094001, 2009.

[8]   M. V. Berry and M. R. Dennis. Knotted and linked phase singularities in monochromatic waves. *Proc. R. Soc. Lond., Ser. A, Math. Phys. Eng. Sci.*, 457(2013):2251–2263, 2001.

[9]   K. Y. Bliokh, A. Y. Bekshaev, and F. Nori. Extraordinary momentum and spin in evanescent waves. *Nat. Commun.*, 5(1):1–8, 2014.

[10] K. Y. Bliokh and F. Nori. Transverse spin of a surface polariton. *Phys. Rev. A*, 85(6):061801, 2012.

[11] K. Y. Bliokh and F. Nori. Transverse and longitudinal angular momenta of light. *Phys. Rep.*, 592:1–38, 2015.

[12] T. Carozzi, R. Karlsson, and J. Bergman. Parameters characterizing electromagnetic wave polarization. *Phys. Rev. E*, 61(2):2024, 2000.

[13] M. R. Dennis, R. P. King, B. Jack, K. O'Holleran, and M. J. Padgett. Isolated optical vortex knots. *Nat. Phys.*, 6(2):118–121, 2010.

[14] R. Dorn, S. Quabis, and G. Leuchs. Sharper focus for a radially polarized light beam. *Phys. Rev. Lett.*, 91(23):233901, 2003.

[15] A. H. Firester, M. E. Heller, and P. Sheng. Knife-edge scanning measurements of subwavelength focused light beams. *Appl. Opt.*, 16(7):1971–1974, 1977.

[16] I. Freund. Cones, spirals, and Möbius strips, in elliptically polarized light. *Opt. Commun.*, 249(1–3):7–22, 2005.

[17] I. Freund. Optical Möbius strips in three-dimensional ellipse fields: I. Lines of circular polarization. *Opt. Commun.*, 283(1):1–15, 2010.

[18] E. J. Galvez, I. Dutta, K. Beach, J. J. Zeosky, J. A. Jones, and B. Khajavi. Multitwist Möbius strips and twisted ribbons in the polarization of paraxial light beams. *Sci. Rep.*, 7(1):1–9, 2017.

[19] T. Grosjean, I. A. Ibrahim, M. A. Suarez, G. W. Burr, M. Mivelle, and D. Charraut. Full vectorial imaging of electromagnetic light at subwavelength scale. *Opt. Express*, 18(6):5809–5824, 2010.

[20] C. Huber, S. Orlov, P. Banzer, and G. Leuchs. Corrections to the knife-edge based reconstruction scheme of tightly focused light beams. *Opt. Express*, 21(21):25069–25076, 2013.

[21] C. Huber, S. Orlov, P. Banzer, and G. Leuchs. Influence of the substrate material on the knife-edge based profiling of tightly focused light beams. *Opt. Express*, 24(8):8214–8227, 2016.

[22] H. Larocque, D. Sugic, D. Mortimer, A. J. Taylor, R. Fickler, R. W. Boyd, M. R. Dennis, and E. Karimi. Reconstructing the topology of optical polarization knots. *Nat. Phys.*, 14(11):1079–1082, 2018.

[23] M. Lax, W. H. Louisell, and W. B. McKnight. From maxwell to paraxial wave optics. *Phys. Rev. A*, 11(4):1365, 1975.

[24] J. Leach, M. R. Dennis, J. Courtial, and M. J. Padgett. Knotted threads of darkness. *Nature*, 432(7014):165, 2004.

[25] K. G. Lee, H. W. Kihm, J. E. Kihm, W. J. Choi, H. Kim, C. Ropers, D. J. Park, Y. C. Yoon, S. B. Choi, D. H. Woo, et al. Vector field microscopic imaging of light. *Nat. Photonics*, 1(1):53–56, 2007.

[26] P. Lodahl, S. Mahmoodian, S. Stobbe, A. Rauschenbeutel, P. Schneeweiss, J. Volz, H. Pichler, and P. Zoller. Chiral quantum optics. *Nature*, 541(7638):473–480, 2017.

[27] P. Marchenko, S. Orlov, C. Huber, P. Banzer, S. Quabis, U. Peschel, and G. Leuchs. Interaction of highly focused vector beams with a metal knife-edge. *Opt. Express*, 19(8):7244–7261, 2011.

[28] M. Neugebauer, T. Bauer, A. Aiello, and P. Banzer. Measuring the transverse spin density of light. *Phys. Rev. Lett.*, 114(6):063901, 2015.

[29] M. Neugebauer, J. S. Eismann, T. Bauer, and P. Banzer. Magnetic and electric transverse spin density of spatially confined light. *Phys. Rev. X*, 8(2):021042, 2018.

[30] L. Novotny and B. Hecht. *Principles of nano-optics*. Cambridge University Press, 2006.

[31] M. Prummer, B. Sick, B. Hecht, and U. P. Wild. Three-dimensional optical polarization tomography of single molecules. *J. Chem. Phys.*, 118(21):9824–9829, 2003.

[32] S. Quabis, R. Dorn, M. Eberler, O. Glöckl, and G. Leuchs. The focus of light–theoretical calculation and experimental tomographic reconstruction. *Appl. Phys. B*, 72(1):109–113, 2001.

[33] B. Richards and E. Wolf. Electromagnetic diffraction in optical systems, ii. Structure of the image field in an aplanatic system. *Proc. R. Soc. Lond., Ser. A, Math. Phys. Eng. Sci.*, 253(1274):358–379, 1959.

[34] N. Rotenberg and L. Kuipers. Mapping nanoscale light fields. *Nat. Photonics*, 8(12):919–926, 2014.

[35] M. B. Schneider and W. W. Webb. Measurement of submicron laser beam radii. *Appl. Opt.*, 20(8):1382–1388, 1981.

[36] C. J. R. Sheppard. Jones and Stokes parameters for polarization in three dimensions. *Phys. Rev. A*, 90(2):023809, 2014.

[37] B. Sick, B. Hecht, and L. Novotny. Orientational imaging of single molecules by annular illumination. *Phys. Rev. Lett.*, 85(21):4482, 2000.

[38] E. H. Synge. Xxxviii. A suggested method for extending microscopic resolution into the ultra-microscopic region. *The London, Edinburgh, and Dublin Philosophical Magazine and Journal of Science*, 6(35):356–362, 1928.

9 Polarization elements that we use in the lab

Polarization elements are important tools for various experiments, from very traditional interferometry to the most recent quantum key distribution experiments. This chapter is focused on the commercial polarization elements that today (2020) one can buy from companies like Thorlabs, Edmund Optics, and Laser Components. Similar to all textbooks, this one will get out of date at some point, but this chapter will be the first to 'expire', because new, more advanced elements will be developed and manufactured. Even now, new 'user-inspired' products appear almost every month, and this rate will certainly get even higher. Nevertheless, it is useful to give a review of what is available right now. We will consider waveplates and rotators for polarization transformations, beam displacers and polarization prisms for the measurement of polarization states, and finally spatial light modulators, elements that use polarization for preparing structured light beams. The few elements of fiber polarization optics will also be briefly reviewed.

9.1 Waveplates

A waveplate (retardation plate) can be most simply made out of a piece of crystalline quartz, cut so that the optic axis (z) is in its plane. Quartz is a positive uniaxial crystal, $n_e > n_o$. Therefore, for light polarized linearly along the z-axis, the refractive index $n = n_e$, and the phase (and group) velocity is smaller than for light polarized orthogonally to z-axis ($n = n_o$). The 'fast axis', perpendicular to z, is sometimes marked by a flat cut.

9.1.1 Multiple-order plates

A plate of reasonable thickness (0.5 mm or thicker) will provide a given phase only for light with a sufficiently narrow bandwidth. Indeed, the phase δ of a plate with the thickness l (see Chapter 5) depends strongly on the wavelength λ,

$$\delta = \frac{\pi \Delta n l}{\lambda}, \tag{9.1}$$

even if the birefringence Δn has no dispersion. For instance, at a wavelength of 600 nm, the birefringence of quartz is $\Delta n = 0.0091$, and at 700 nm, it is $\Delta n = 0.0090$, nearly the same over 100 nm. Then for the wavelength $\lambda = 600$ nm the phase of a plate with thickness $l = 0.495$ mm will be, according to Eq. (9.1), 7.5π. This means that the plate will be a HWP for this wavelength, because the phase delay multiple to π will not affect the polarization transformation. The phase delay that matters for this plate is $\pi/2$. Such a plate is called a multiple-order plate. More precisely, a waveplate with

https://doi.org/10.1515/9783110668025-009

the phase $\delta = (m + 1/2)\pi$ is a HWP of order m. The plate we are considering is therefore a 7th-order HWP.

At the same time, by using Eq. (9.1) again we find that for the wavelength $\lambda' = 615$ nm the same plate will produce a phase delay of $\delta' = 7.25\pi$ and will then be a QWP of order 7. For intermediate wavelengths, $600 < \lambda < 615$ nm, the plate will be neither a HWP nor a QWP, but something in between.

We see that a multiple-order waveplate can be only used for very narrowband radiation. For instance, if we request that the deviation from the expected phase shift should be not more than 0.1 rad, the plate considered here will be suitable for a bandwidth of only about 2 nm.

9.1.2 Zero-order plates

For a zero-order plate, the phase delay should be $\delta \leq \pi/2$: exactly $\pi/2$ for a HWP and exactly $\pi/4$ for a QWP. For the wavelength $\lambda = 600$ nm, a zero-order HWP should have a thickness of 33 μm, and a zero-order QWP, a thickness of 16.5 μm. Such plates will have a broader bandwidth but they will be too fragile. To avoid this, one can make a zero-order plate by stacking together two plates, with the optic axes orthogonal and thicknesses l_1, l_2 large enough to provide the necessary rigidity (Fig. 9.1). Because the extraordinary beam in the first plate will be the ordinary beam in the second plate and vice versa, the phase delays in the two plates will be of different sign. The resulting phase delay will be

$$\delta = \frac{\pi \Delta n (l_1 - l_2)}{\lambda}. \tag{9.2}$$

It follows that this composite plate has an effective thickness $l_1 - l_2$. If l_1 and l_2 are close enough, then the composite plate will act like a zero-order plate. For instance, a plate with $l_1 = 533$ μm and $l_2 = 500$ μm will be a zero-order HWP for $\lambda = 600$ nm.

A composite zero-order waveplate can be made by gluing two quartz plates together—in this case it does not stand very strong radiation because the glue can

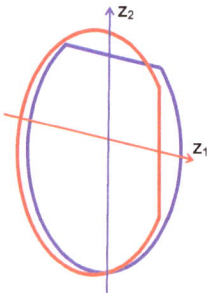

Figure 9.1: A zero-order waveplate made of two pieces of crystalline quartz.

burn. To get a higher damage threshold, the two plates can be simply stacked together using the 'optical contact'.

There are also so-called 'true zero-order plates'. Such a plate is a thin layer of polymer (liquid crystal) placed between two glass plates. The liquid crystal provides in this case birefringence while the isotropic glass plates make the whole construction mechanically strong.

9.1.3 Dual-wavelength plates

The thickness of a plate can be chosen such that the plate will perform different polarization transformations for different wavelengths, for instance, it can be a HWP for one wavelength and a QWP for another one. An example was discussed in Section 9.1.1: a plate of thickness 495 µm will be a HWP for the wavelength $\lambda' = 600$ nm but a QWP for the wavelength $\lambda' = 615$ nm. Alternatively, a HWP for a certain wavelength can simultaneously have $\delta = m\pi$ for another one, then it will not affect the polarization of light at this other wavelength.

9.1.4 Achromatic waveplates

Even for a zero-order plate, the bandwidth is not very large. In the example considered in Section 9.1.2, a zero-order plate for $\lambda = 600$ nm, with the (effective) thickness 33 µm, will have a bandwidth of 30 nm. Clearly, the bandwidth of a plate scales as the inverse of its thickness.

One can increase this bandwidth by compensating the wavelength in the denominator of Eq. (9.2) with the wavelength-dependent birefringence Δn in its numerator. In the normal dispersion range, the birefringence reduces with the wavelength and such compensation is impossible. To overcome this problem, the two plates forming a zero-order plate are made of different crystals, for instance, quartz and magnesium fluoride (MgF_2) or quartz and sapphire (Al_2O_3) [6]. Then, instead of Eq. (9.2), the phase of the plate will be given by

$$\delta = \frac{\pi(\Delta n_1 l_1 - \Delta n_2 l_2)}{\lambda}, \tag{9.3}$$

where the indices 1, 2 correspond to the two different materials. Because the two materials have different dependence of the birefringence on the wavelength, by properly choosing the lengths l_1 and l_2 the numerator in this expression can be made roughly scaling with the wavelength. Such a waveplate will have the same phase within a very large bandwidth, up to about 200 nm.

9.1.5 Variable waveplates

Some experiments require waveplates with a variable phase. As such a *variable wave-plate*, it is convenient to use a Soleil-Babinet compensator (Fig. 9.2). It consists of a plate with a fixed thickness and another one, with a variable thickness. Both plates are usually made of quartz. The optic axes of the two plates are orthogonal (shown by blue dots and arrow), as in the case of a zero-order plate. To make the thickness of the first plate variable, the plate is made out of two wedges, which can be displaced with respect to each other (shown by a gray arrow in the figure). The phase of such a plate can be varied from 0 to π. Due to the use of two plates with orthogonal optic axes, the Soleil-Babinet compensator is equivalent to a zero-order plate.

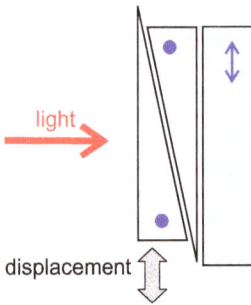

Figure 9.2: Soleil-Babinet compensator as a variable waveplate. The optic axis directions are shown by blue dots and arrow.

9.2 Rotators

In many devices or experiments, it is necessary to rotate linear polarization by a certain angle regardless of its initial state. In such cases, as discussed in Chapter 5, a good solution is a polarization rotator.

In the simplest case, a polarization rotator is a plate of crystalline quartz, cut orthogonal to the optic axis. As described in Chapter 4, such a *z-cut* has circularly polarized normal waves and circular birefringence. For light in the visible range, a slab with a few mm thickness will rotate the polarization by 90° (Fig. 5.2). The disadvantage of such a rotator is its small bandwidth. For instance, a 2 mm slab will rotate the polarization by 90° for light at 600 nm but by 45° at 700 nm.

For operation with different wavelengths, much more convenient are Faraday rotators. These are rotators based on the Faraday effect: in a longitudinal magnetic field H, the plane of polarization rotates by angle $\delta = VHL$, where L is the length of the nonlinear crystal and V is the Verdet constant [2]. The Verdet constant is determined by the nonlinear properties of the material and therefore depends on the wavelength. But for any wavelength it is possible to set the magnetic field in such a way that a given rotation angle δ is achieved. Faraday rotators are made of ferromagnetic materials like terbium-gallium-garnet [6], combined with permanent magnets, and operate within a bandwidth of 50–100 nm.

An important difference between a Faraday rotator and a rotator based on materials with natural optical activity (like crystalline quartz) is that the former performs a *non-reciprocal polarization transformation*. In a Faraday rotator the direction of polarization plane rotation does not depend on the light propagation direction, but only on the direction of the magnetic field. If a beam is sent through a Faraday rotator and then reflected back, the angles of rotation acquired on both paths add up instead of canceling each other, as in the case of a material with natural optical activity.

This effect is used in devices known as *optical isolators*, which enable 'one-way' propagation of light and are therefore used to eliminate back reflections. An optical isolator contains a Faraday rotator with a magnet and polarizers placed before and after it (Fig. 9.3). Consider, for example, the case where light at the input (on the left) is linearly polarized so that it passes through the first polarizer, which transmits horizontally polarized light. The magnetic field is such that the Faraday rotator rotates the polarization clockwise by 45°. The second polarizer is oriented so that it transmits the resulting diagonally polarized light. The incident light is therefore fully transmitted. On the way back, light propagates from right to left; then at the entrance of the Faraday rotator it will be initially diagonally polarized. After the Faraday cell, its polarization plane will further rotate by 45°. Then it will become vertically polarized and will not pass through the first polarizer.

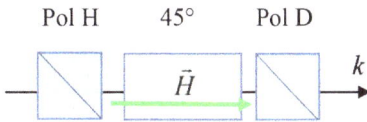

Figure 9.3: An optical isolator based on a Faraday rotator.

9.3 Beam displacers

A beam displacer (Fig. 9.4) is a birefringent crystal (usually calcite, but sometimes also α-barium borate (BBO), magnesium fluoride, or quartz [6]) cut in such a way that the optic axis is at 45° to the normal to the front face. Due to the spatial walk-off, an incident beam in an arbitrarily polarization state will split inside the calcite crystal in two beams. The ordinary beam, polarized orthogonally to the principal plane (the plane containing the wave vector and the optic axis), will have the Poynting vector parallel to the wave vector. The extraordinary beam, polarized in the principal plane, will have the Poynting vector tilted with respect to the wave vector by the walk-off angle. In the case of calcite, for instance, the walk-off angle for the visible spectral range is as large as 6°. At the output of the beam displacer, the ordinary and extraordinary beams are again parallel to each other, as there is no anisotropy any more and the Poynting vector is again parallel to the wavevector. But the extraordinary beam will be displaced.

Calcite with the optic axis at 45°

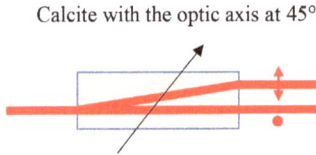

Figure 9.4: A beam displacer.

Beam displacers can be useful in many optics experiments. For instance, with the help of a beam displacer one can realize Young's double-slit interference: at the output of the displacer the ordinary and extraordinary beams represent two copies of the same beam. For the interference to take place, both beams should have the same polarization state, but this can be achieved by placing another polarizer after the displacer, projecting both polarization states on a single one. In other types of interferometers, beam splitting can be also conveniently realized with the help of several beam displacers. As an example, Fig. 9.5 shows a Mach–Zehnder interferometer formed by two beam displacers with the optic axes parallel. A HWP oriented at 45°, placed between them, converts the ordinary beam at the output of the first displacer into the extraordinary beam in the second one, and vice versa. Such an interferometer is easy to align and it does not suffer from rapid phase drift. By placing more beam displacers, the interferometer can be transformed into a multipath one.

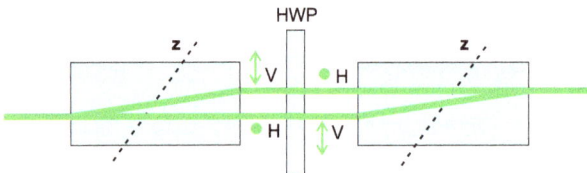

Figure 9.5: A Mach–Zehnder interferometer formed by two beam displacers and a HWP.

Even more important is the use of beam displacers in the Stokes measurement, already discussed in Chapter 4. With a camera registering both beams after a beam displacer (see Fig. 4.8), one can measure the intensities for the horizontally and vertically polarized radiation and this way measure the Stokes variables S_1 and S_0. For the measurement of S_2, the displacer should be rotated 45° and for the measurement of S_3, it should be preceded by a QWP. Especially convenient is that by varying the size of the incident beam, one can modify the quality of the measurement. If the beam is broader than the transverse displacement between the ordinary and extraordinary beams, the measurement becomes 'weak'. This might seem to be a bad idea, but for the reasons that will be clear further, weak measurements are widely used in quantum optics. They will be considered in detail in Chapter 11.

9.4 Prisms and polarizing beamsplitters

Polarization prisms have been already mentioned in this book several times, because they are one of the main elements of polarization optics. Their role is to separate light with orthogonal linear polarizations in two beams. In this section we will consider different modifications of such prisms.

Most well-known are various versions of Glan prisms. Their operation is based on the total internal reflection as well as on the Brewster effect. Let us briefly consider both effects here; for a more detailed consideration, see [1].

9.4.1 Total internal reflection and Brewster's law

Total internal reflection follows from the Snell law: for the angles of incidence and refraction θ_i, θ_r at an interface between the two materials, whose refractive indices are n_1 and n_2 (Fig. 9.6) [1],

$$\frac{\sin \theta_i}{\sin \theta_r} = \frac{n_2}{n_1}. \tag{9.4}$$

If the second medium is less dense (for instance, air), then $n_1 > n_2$, and total internal reflection is possible. The angle θ_{i0} of total internal reflection is defined as

$$\sin \theta_{i0} = \frac{n_2}{n_1}. \tag{9.5}$$

For a beam incident at this angle, $\theta_r = \pi/2$, and the refracted beam propagates along the interface. For larger angles of incidence, total internal reflection occurs and there is no refracted beam.

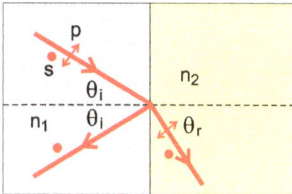

Figure 9.6: Refraction and reflection of a beam at an interface.

The Brewster law follows from the Fresnel formulas for reflectivity and transmissivity of an interface. If the angle of incidence is nonzero, it is convenient to introduce the notation 'p' and 's', correspondingly, for the waves polarized parallel to the plane of incidence (parallel) and orthogonal to it (from the German *senkrecht*). From the Fresnel formulas, it follows [1] that reflectivity and transmissivity differ for p- and s-polarized beams. In particular, for a beam incident at the *Brewster angle*, defined as

$$\tan \theta_B = \frac{n_2}{n_1}, \tag{9.6}$$

the p-component is not reflected at all. The reason for this can be qualitatively understood if we notice that condition (9.6) means that the reflected and refracted beams are orthogonal to each other (see Fig. 9.6). Indeed, from Snell's law (9.4) in combination with Eq. (9.6), we obtain $\sin\theta_i = \cos\theta_r$. Then the absence of the 'p' polarization in the reflected beam can be explained in simple terms. Recall that the reflected beam emerges due to the oscillation of electrons on the right of the interface. Then, for the reflected beam to be 'p'-polarized, the electrons should oscillate in the transverse direction. But this is exactly the direction in which the refracted beam propagates; therefore, these oscillations cannot be excited by the refracted beam.

Unlike total internal reflection, which occurs only at the boundary with a less dense medium, the Brewster effect does not require this condition. The relation between the two refractive indices n_1, n_2 only affects the value of the Brewster angle.

The Brewster effect is used for making so-called Brewster windows. These are plates made of a crystalline or glass material and placed at the Brewster angle, so that they do not reflect an incident p-polarized beam. As a result, there is no loss for the p-polarization. Such windows are used for gas laser tubes, in which they reduce the losses. Brewster-angle incidence is also used in all kinds of prisms, to reduce reflection losses.

9.4.2 Glan–Taylor prism

Figure 9.7 shows a Glan–Taylor prism. Its two halves, separated by an air gap, are made of crystalline material. Usually calcite is chosen, due to its large birefringence and small dispersion. For the operation in the UV range, Glan–Taylor prisms can be also made of α-BBO [6], which has lower birefringence than calcite but is transparent down to 200 nm. The optic axes in both halves of the prisms are vertical (shown by blue arrows in the figure).

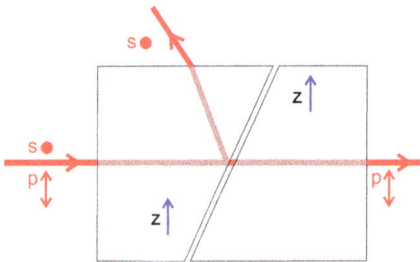

Figure 9.7: A Glan–Taylor prism.

A light beam incident on the prism and then hitting the air gap contains, in general, both s- and p-polarized components. The s-component is an ordinary beam whose refractive index is n_o, while the p-component (extraordinary beam) has the refractive index n_e. These two indices differ considerably; for instance, for calcite, at 532 nm the values of n_o and n_e are, respectively, 1.66 and 1.49. The angle of incidence on the crystal-air interface is chosen to be larger than the total internal reflection angle for the s-polarized beam: $\sin \theta_i > 1/n_o$. Therefore, the s-polarized beam is fully reflected from the interface and the transmitted beam is perfectly cleaned from the s-polarization (horizontal in the figure). At the same time, for the p-polarized (extraordinary) beam, the angle of incidence at the same interface is equal to the Brewster angle: $\tan \theta_i = 1/n_e$. Therefore, no p-polarization (vertical) is contained in the reflected beam. Typically, a Glan–Taylor prism provides an extinction ratio of 1000000 : 1, which means that the transmissivity for the incident s-polarized beam is six orders of magnitude lower than for the p-polarized beam. For the reflected beam, the extinction ratio is somewhat worse but still high enough.

Because the angle of incidence is only a bit larger than the total internal reflection angle, a Glan–Taylor prism does not provide perfect polarization splitting for divergent beams. It is very important therefore that the incident beam is collimated and hits the prism at normal incidence.

A Glan–Laser prism has the same design as the Glan–Taylor prism but it is made of high-quality crystal material and therefore stands a higher intensity and produces less scattered light.

9.4.3 Glan–Thompson prism

A Glan–Thompson prism has a slightly different structure (Fig. 9.8). The angle of incidence on the interface is larger, and in both halves of the prism, the optic axes (z) are parallel to the interface (blue dots in the figure). Now the s and p components change roles: the first one is an extraordinary beam and is transmitted (with low loss due to the Brewster law), and the second one is an ordinary beam and is reflected.

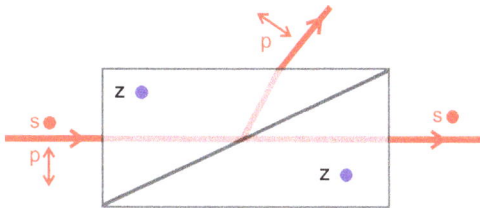

Figure 9.8: A Glan–Thompson prism.

Because of the larger angle of incidence, the gap between the two halves can be filled with optical cement instead of the air. This reduces the tolerance of the prism to damage but increases the range of angles for which it can operate. The width of the field of view for a Glan–Thompson prism can be as large as 40° [6].

In a Glan–Thompson prism, unlike in the Glan–Taylor or Glan–Laser prisms, the reflected beam is usually blocked. The prism is used therefore only as a polarizer and not as a beam splitter.

9.4.4 Wollaston prism

For a Wollaston prism (Fig. 9.9), similarly to Glan prisms, two halves are made of bire-fringent crystalline material, which can be calcite, α-BBO, quartz, or magnesium fluo-ride. But in contrast to the Glan prisms, there is no gap between the two halves, and the refractive indices for an incident beam differ because the optic axes in the two halves are orthogonal to each other. For instance, in Fig. 9.9, the input half of the prism has the optic axis horizontal and the output part, vertical. Then a horizontally polarized beam is an extraordinary polarized beam in the first half, and has a refractive index n_e, but it is an ordinary beam after the interface and its refractive index is n_o. For an incident beam with vertical polarization, it is the other way round. In the case of cal-cite, which is a negative crystal, $n_o > n_e$. Therefore, at the boundary the horizontally polarized beam goes from a less dense medium into a more dense one; this beam is re-fracted downwards. The vertically polarized beam, on the contrary, enters a less dense medium and is therefore refracted upwards. The angular separation of the beams can be different, from 1° to 20°, depending on the orientation of the interface.

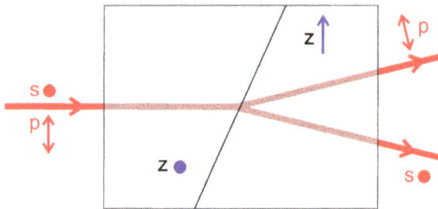

Figure 9.9: A Wollaston prism.

Calcite Wollaston prisms provide an extinction ratio as high as 1000000 : 1 for both vertically and horizontally polarized beams. For this reason, a Wollaston prism is the best option if both beams are to be used after the splitting.

9.4.5 Polarizing cubes and plates

The simplest and cheapest polarizers are made with the help of dielectric coatings placed on isotropic materials, like glass or fused silica. For instance, a beam incident at 45° on a polarizing plate will experience high transmissivity for the p-polarization and high reflectivity for the s-polarization. In order to eliminate the reflection from the second surface of the plate, the plate is made slightly wedged. The extinction ratio for a polarizing plate can be up to 10000 : 1, considerably worse than for Glan or Wollaston prisms.

Polarizing cubes are based on a similar principle as polarizing plates, but they are easier to align. A polarizing cube is made of an isotropic material with a dielectric coating covering the 45° interface between its two halves. The dielectric coating is chosen such that the p-polarization is not reflected due to the Brewster effect. The s-polarization is partly reflected, partly transmitted, but it can be made to be reflected completely due to the interference. The extinction ratio of a polarizing cube is typically 1000 : 1, worse than for a polarizing plate.

9.5 Fiber polarization components

Fiber optics is an extremely important part of modern optics, and there is hardly any laboratory not using fiber components. Therefore, although polarization of light is not as crucial for fibers as it is for free-space propagating beams, here we will shortly describe fiber components dealing with polarization.

9.5.1 Polarization maintaining fiber

In a standard single-mode fiber, polarization state of light is not maintained. This makes usual fibers inconvenient for many experiments. To overcome this problem, polarization-maintaining (PM) fibers have been designed. These fibers have a core that is not axially symmetric. To achieve this, the core either has elliptical cross-section or is accompanied by additional features inducing transverse stress and thus violating the axial symmetry (bow-tie fibers, Panda fibers etc.) Such a fiber features birefringence: the refractive index is different for light polarized along two orthogonal directions of the core symmetry. Similarly to the case of a waveplate, these directions are usually called 'fast' and 'slow' axes.

A PM fiber, similar to a waveplate, has eigenmodes polarized linearly. But, in contrast to a quartz waveplate, the birefringence of such a fiber is tiny, on the order of 10^{-3}–10^{-4}. Typically a PM fiber is characterized by its beat length, which is defined as the length needed to introduce a phase $\delta = \pi$.[1] The beat length of a fiber usually scales with the wavelength and amounts to a few millimeters.

But the most important characteristic of a PM fiber is its H-parameter, which indicates namely how well the fiber maintains the input linear polarization. There is a certain limit to the length of a PM fiber because of the inevitable cross-talk between the two polarization modes it supports. The H-parameter is defined as the extinction

---

1 A beat length of a PM fiber is equivalent to twice the thickness of a zero-order HWP.

ratio per meter. For instance, if the H-parameter is $10^{-6}\,\text{m}^{-1}$, then after 1 km of this fiber its extinction of the 'wrong' polarization is 30 dB.

With the help of a PM fiber, fiber polarization beam splitters are produced. In such a device, a single-mode fiber is connected to the input of a polarization prism (for instance, made of calcite), whose two outputs are, in turn, coupled to PM fibers with fast axes oriented orthogonally. Such fiber splitters are used for splitting orthogonally polarized modes of a fiber, or combining them (in the latter case the device is used in the opposite direction).

9.5.2 Fiber polarization controllers

To control the polarization of light after or before an optical fiber, there are special devices called fiber paddles. In each such paddle, a fiber is several times looped around a spool, which leads to a birefringence appearing in the fiber, the fast axis being in the plane of the loops. The phase introduced by such a paddle is [6]

$$\delta = \frac{2\pi^2 a N d^2}{\lambda D},\tag{9.7}$$

where a is a constant depending on the material of the fiber, N the number of loops, d the cladding diameter, λ the wavelength, and D the diameter of the loop. By properly choosing the diameter of a single loop and the number of loops, one can make a paddle equivalent to a QWP or to a HWP. A system of three paddles, two HWPs and a QWP between them, provides an arbitrary polarization transformation from any input state to any desired output state. The polarization controller is operated by changing the tilt of each of the three paddles.

9.6 Liquid-crystal devices

Section 4.2.4 briefly described the polarization properties of liquid crystals. We saw that a nematic liquid crystal behaves as a uniaxial crystal whose optic axis direction can be oriented by external electric field. In this section, we will consider some optical devices based on this property. Such devices are now ubiquitous in our everyday life, and certainly in scientific laboratories. In particular, they are key elements to prepare scalar and vector structured beams, whose overview was given in Chapter 7.

9.6.1 Spatial light modulators

Spatial light modulators (SLMs) are used in numerous fields of optics, from holography to beam shaping, and in standard devices like beam projectors. An SLM is a device for modulating the phase of a beam, and sometimes also its amplitude, with high spatial resolution. The operation of an SLM is based on the polarization properties of liquid crystals (Chapter 4). This operation is schematically demonstrated in Fig. 9.10 [4] where, as an example, four pixels of an SLM are shown.

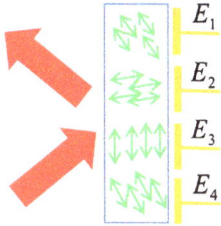

Figure 9.10: Operation of a spatial light modulator. Green arrows show the directors of the molecules and red arrows, the incident and reflected light beams.

Each pixel contains a tiny amount of a liquid crystal. Molecules of the liquid crystal (their directors shown by green arrows) are oriented by the electric field E_i, applied independently to every pixel. As a result, in every pixel there is a different orientation of all molecules. As described in Section 4.2.4, the refractive index for light with a certain polarization will be different depending on the orientation of the molecules in each pixel. Light (shown by red arrows) is sent to the SLM and reflected by a mirror on its back side. The phase acquired by the incident light at each pixel will be proportional to the refractive index value, and it can be varied independently by setting the electric field at the pixel. The accessible phases cover the range from 0 to 2π.

This way a different phase can be imparted on every pixel. To make sure that only phase-modulated beam is used, usually a blazed grating is also written on the SLM. The superposition of the blazed grating and the required phase distribution defines the phase profile of the diffracted beam.

As mentioned, one of the applications of an SLM is generation of scalar structured beams. For example, to convert a Gaussian beam into a Hermite–Gaussian or a Laguerre–Gaussian beam, one has to modulate both its phase and its intensity as in the examples of Fig. 7.2. The phase distributions, shown in the insets of the figure, are then imparted on the beam with the help of an SLM. For modulating the intensity, additional elements have to be used. In particular, because a liquid crystal modifies not only the phase but also the polarization state of a beam, the intensity can be modulated by means of a polarizer introduced after the SLM.

9.6.2 Q-plates

The fact that a liquid crystal modifies the polarization state of light underpins the operation of so-called q-plates—devices that appeared recently but are already available commercially [5]. A q-plate is designed to substitute segmented waveplates mentioned in Chapter 7 as well as more complicated devices. Indeed, a liquid crystal layer with all molecules oriented the same way is similar to a slab of a birefringent crystal. Suppose that the directors of all molecules are parallel and lie in the plane of the layer. Then, depending on the thickness of the layer, the liquid crystal will act as a waveplate with a certain phase, for instance a HWP or a QWP. A segmented waveplate can then be made by orienting the directors of the molecules differently in different segments of such a layer.

Figure 9.11 shows, as an example, a 8-segment waveplate made of a liquid crystal. In each next segment of this waveplate, the directors of the molecules are rotated by $\pi/8$ compared to the previous one (panel a). This construction has been also described in Section 7.5. Let the thickness of the liquid be such that each segment acts as a HWP, with the optic axis along the directors. Then, if the input light beam is horizontally polarized, its polarization after the kth segment will be still linear but rotated by $(k - 1)\pi/4$. Panel b illustrates this transformation as a rotation on the Poincaré sphere, with the numbered green lines marking the rotation trajectories. Segments 1 and 5 lead to no rotation. Panel d below shows the resulting distribution of the polarization after the plate. The output beam is radially polarized, similar to the one discussed in Chapter 7.

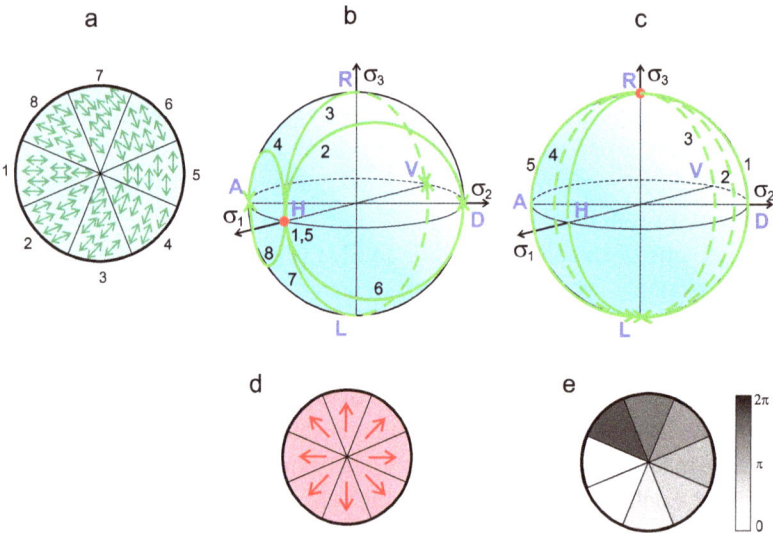

Figure 9.11: An 8-segment q-plate (a) with the thickness equivalent to the one of a HWP. Green arrows show the directors of the molecules. Panels b, c show the Poincaré-sphere representation of the q-plate's action on a horizontally polarized beam (b) and a right-hand circularly polarized beam (c). Red points show the Stokes vector of the initial polarization state and the numbered green lines are rotations performed on the Poincaré sphere by different segments of the plate. After the q-plate, a horizontally polarized beam becomes a radially polarized beam (d) and a right-hand circularly polarized beam becomes left-hand circularly polarized and acquires an azimuthal phase shift (e).

If the input beam is right-hand circularly polarized, we know from Chapter 5 that after a HWP it will be left-hand circularly polarized, regardless of the plate orientation. But, as described in Section 6.4.2 of Chapter 6, depending on the plate orientation it will acquire a different geometric phase (see Figure 6.9). Figure 9.11(c) shows the Poincaré-sphere representation of the polarization transformations in this case. For instance, segment 1 of the plate results in the rotation of the initial Stokes vector (red point on the North Pole) by π around the σ_1 axis. Segment 2 rotates the same point by the same angle around the axis $(\sigma_1 + \sigma_2)/\sqrt{2}$ axis, segment 3, around the σ_2 axis and so on. After segment k, the solid angle $\Omega_k = (k-1)\pi/2$ is covered on the Poincaré sphere compared to the trajectory due to the first segment, and a geometric phase $\beta_k = (k-1)\pi/4$ is acquired. The resulting beam will have a phase varying with the azimuthal angle by 2π, i. e., an orbital angular momentum $l = 1$ (panel e).

In a similar way, one can azimuthally modulate both the phase and the polarization state of the beam. For instance, if the waveplate is a QWP, and the input beam is right-hand circularly polarized, then the output beam will have radial polarization and an azimuthally varying phase. This becomes clear from Fig. 9.11(c) if we imagine the trajectories stopped half-way, according to the fact that a QWP performs only a $\pi/2$ rotation on the Poincaré sphere.

In order to make a q-plate [5], a nematic liquid crystal is placed between two polymer-coated substrates. The polymer is corrugated or structured using polarized ultraviolet light; then the directors of the nematic liquid crystal are aligned accordingly. This way, any spatial distribution of the molecules' directors can be obtained. In particular, the boundaries between the segments in Fig. 9.11(a) can be made smooth, so that the polarization state of light or a phase at the output of a q-plate is varied continuously and not stepwise as in panels d and e of the figure.

Some q-plates enable tuning of their phase delay at a given point by means of applied AC electric field, which changes the birefringence of the oriented liquid crystal. Moreover, it is possible to independently control the phases for different wavelengths. This is a considerable advantage compared to a segmented waveplate [3].

Bibliography

[1] M. Born and E. Wolf. *Principles of optics*. Pergamon Press, 1970.
[2] D. Goldstein. *Polarized light*. GRC, 2003.
[3] A. Rubano, F. Cardano, B. Piccirillo, and L. Marrucci. Q-plate technology: a progress review. *J. Opt. Soc. Am. B*, 36(5):D70–D87, May 2019.
[4] Hamamatsu selection guide. https://www.hamamatsu.com/resources/pdf.
[5] ARCoptics website. http://www.arcoptix.com/Q_Plate.htm.
[6] Thorlabs website. https://www.thorlabs.com.

10 Polarization in nonlinear optics

Nonlinear optics describes frequency conversion of light due to the nonlinearity of the electromagnetic response of the matter. Many aspects of nonlinear optics are considerably based on the polarization of light. This is why this chapter is included into this book. However, the reader is expected to have some basic knowledge of nonlinear optics.

10.1 Nonlinear susceptibilities: tensor description

In Chapter 2, the polarization of the matter $\vec{P}(\vec{r}, t)$ was introduced as the dipole moment induced in a unit volume by an external electric field $\vec{E}(\vec{r}, t)$. In Chapter 4, the relation between the polarization $\vec{P}(\vec{r}, t)$ (and therefore the displacement $\vec{D}(\vec{r}, t) = \epsilon_0 \vec{E}(\vec{r}, t) + \vec{P}(\vec{r}, t)$) and the electric field was assumed to be linear, see Eq. (4.5), because at that point we were only interested in the tensor properties of the linear permittivity and susceptibility. Now we will consider a more general picture. It is well known [2] that due to the anharmonicity of the matter, the polarization depends on the electric field in a nonlinear fashion,

$$\vec{P}(\vec{r}, t) = \epsilon_0 [\hat{\chi}^{(1)} \cdot \vec{E}(\vec{r}, t) + \hat{\chi}^{(2)} : \vec{E}(\vec{r}, t)\vec{E}(\vec{r}, t)$$

$$+ \hat{\chi}^{(3)} \vdots \vec{E}(\vec{r}, t)\vec{E}(\vec{r}, t)\vec{E}(\vec{r}, t) + \cdots], \tag{10.1}$$

where we introduced the nth-order nonlinear susceptibility $\hat{\chi}^{(n)}$. The nonlinear susceptibility $\hat{\chi}^{(n)}$ sets the relation between a vector ($\vec{P}$) and a product of n other vectors ($\vec{E}$) and is therefore a tensor of rank $n + 1$. The tensor nature of the nonlinear susceptibility is often ignored in simplified descriptions of nonlinear optical interactions; but in this section, it will be the main issue. The signs '·', ':', etc. denote the procedure of multiplying a tensor by one, two, etc. vectors. In what follows, we will omit the 'hats' over the susceptibilities but will bear in mind that they are tensors.

We have assumed here that the response of the medium is instantaneous and local: the space and time dependences of the polarization $\vec{P}(\vec{r}, t)$ repeat those of the field $\vec{E}(\vec{r}, t)$. This is possible only in a material without absorption and dispersion [2], but in this book, as a rule, we ignore both effects.

We will further distinguish between the linear polarization $\vec{P}^{(1)}(\vec{r}, t) = \epsilon_0 \chi^{(1)} \cdot \vec{E}(\vec{r}, t)$, second-order nonlinear polarization

$$\vec{P}^{(2)}(\vec{r}, t) = \epsilon_0 \chi^{(2)} : \vec{E}(\vec{r}, t)\vec{E}(\vec{r}, t), \tag{10.2}$$

third-order nonlinear polarization

$$\vec{P}^{(3)}(\vec{r}, t) = \epsilon_0 \chi^{(3)} \vdots \vec{E}(\vec{r}, t)\vec{E}(\vec{r}, t)\vec{E}(\vec{r}, t), \tag{10.3}$$

and so on.

https://doi.org/10.1515/9783110668025-010

Alternatively, Eq. (10.1) can be written using subscript notation:

$$P_i = \epsilon_0 \left(\sum_j \chi_{ij}^{(1)} E_j + \sum_{jk} \chi_{ijk}^{(2)} E_j E_k + \sum_{jkl} \chi_{ijkl}^{(3)} E_j E_k E_l + \cdots \right), \tag{10.4}$$

where we have omitted, for brevity, the time and space dependences of the fields. The indices i, j, k, l correspond to the Cartesian coordinates and each one can take values x, y, z.

In Chapter 4 we have already dealt with the second-rank first-order susceptibility tensor $\chi_{ij}^{(1)}$ and the dielectric permittivity tensor ϵ_{ij} related to it. In the scalar form, we had (see Chapters 2, 4) $\epsilon = 1 + \chi$. In the matrix form, we get for the linear dielectric permittivity tensor

$$\epsilon_{ij} = \delta_{ij} + \chi_{ij}^{(1)}, \tag{10.5}$$

where the Kronecker delta δ_{ij} denotes the unity matrix.

We already know that the dielectric permittivity tensor is symmetric, hence the first-order linear susceptibility tensor is also symmetric,

$$\chi_{ij}^{(1)} = \chi_{ji}^{(1)}. \tag{10.6}$$

From Chapter 4, we know that by properly choosing the frame of reference, the number of elements for the dielectric permittivity tensor can be reduced. The optimal frame of reference corresponds to the symmetry of the matter, and in this frame the maximal number of ϵ_{ij} elements is three. In materials of higher symmetry, it can be two (uniaxial crystals) or one (isotropic crystals or amorphous materials).

In the next two sections, we will consider the tensor properties of nonlinear susceptibilities and, in particular, the number of their independent elements.

10.1.1 Second-order susceptibility

The second-order susceptibility $\chi_{ijk}^{(2)}$ is nonzero only for materials (crystals) without inversion symmetry. This directly follows from Eq. (10.2): suppose that we perform the inversion transformation, i. e., change all coordinates $\vec{r}$ to $-\vec{r}$. Then all vectors should change their signs, but if the material is centrosymmetric, then $\chi_{ijk}^{(2)}$ will stay invariant. Then the left-hand side of Eq. (10.2) will change sign but its right-hand side will remain the same. The only way to resolve this controversy is to claim that in a centrosymmetric material, $\chi_{ijk}^{(2)} = 0$.[1]

---

[1] At the boundary between two isotropic materials, for instance, on a surface of an isotropic solid, one can still observe second-order nonlinear effects because the boundary breaks the central symmetry. In other words, there is always a *surface contribution* to the second-order nonlinear susceptibility.

Usually, nonlinear effects are observed for rather narrowband (laser) radiation, so that one can speak about radiation at certain frequencies converted into radiation at other frequencies. For instance, in *second-harmonic generation* the *fundamental radiation* at a frequency ω is converted into its *second harmonic* at frequency 2ω. It makes then sense to consider the fields and polarization of the matter at specific frequencies, rather than at specific times, as in Eq. (10.1). In other words, we will further use the Fourier components of the fields and nonlinear polarization from Eq. (10.1), and the corresponding Fourier components of the susceptibilities.

Consider now the general case of second-order frequency conversion, for instance, conversion from frequencies ω_n and ω_m (which could be any, including negative values) to the frequency $\omega_n + \omega_m$. This includes sum- and difference-frequency generation, second-harmonic generation, and optical rectification. The second-order polarization at frequency $\omega_n + \omega_m$ is [2]

$$P_i^{(2)}(\omega_n + \omega_m) = \epsilon_0 \sum_{jk} \sum_{nm} \chi_{ijk}^{(2)}(\omega_n + \omega_m, \omega_n, \omega_m) E_j(\omega_n) E_k(\omega_m), \tag{10.7}$$

where we stressed that the value of the susceptibility depends on the process we consider.

For instance, second-harmonic generation will correspond to $\chi_{ijk}^{(2)}(2\omega, \omega, \omega)$; difference-frequency generation, to $\chi_{ijk}^{(2)}(\omega_1 - \omega_2, \omega_1, -\omega_2)$; optical rectification, to $\chi_{ijk}^{(2)}(0, \omega, -\omega)$.

Let us consider the most general case of conversion between three frequencies ω_1, ω_2, and $\omega_1 + \omega_2 \equiv \omega_3$, and ask the question: how many elements of the $\chi_{ijk}^{(2)}(\omega_3, \omega_1, \omega_2)$ tensor are there?

Without any restrictions, there would be six different tensors: $\chi_{ijk}^{(2)}(\omega_1, -\omega_2, \omega_3)$; $\chi_{ijk}^{(2)}(\omega_1, \omega_3, -\omega_2); \ldots; \chi_{ijk}^{(2)}(\omega_3, \omega_1, \omega_2)$. Another six tensors can be obtained by flipping the signs of all frequencies. Finally, in every tensor there will be 27 permutations of indices i, j, k. Altogether, it makes $12 \times 27 = 324$ complex numbers, because susceptibilities are, in the general case, complex.

Fortunately, there are some restrictions imposed by symmetry considerations.

1. Complex conjugation. Because all fields entering Eq. (10.1) are real, the Fourier components of the susceptibility satisfy the condition

$$\chi_{ijk}^{(2)}(-\omega_3, -\omega_1, -\omega_2) = \left[\chi_{ijk}^{(2)}(\omega_3, \omega_1, \omega_2)\right]^*, \tag{10.8}$$

i. e., flipping the signs of the frequencies is equivalent to complex conjugation. This reduces the number of independent tensor values by a factor of 2.

2. Intrinsic permutation symmetry. From the definition of the process (10.7), it follows that the frequencies $\omega_{1,2}$ summing up to a frequency ω_3 can be interchanged together with their indices [2]:

$$\chi_{ijk}^{(2)}(\omega_3, \omega_1, \omega_2) = \chi_{ikj}^{(2)}(\omega_3, \omega_2, \omega_1). \tag{10.9}$$

3. Full permutation symmetry [2]. In a lossless medium there are additional symmetry restrictions imposed on the nonlinear susceptibility tensor. First, in the absence of absorption the imaginary parts of all susceptibilities should be zero. Then $\chi_{ijk}^{(2)}$ should be a real tensor, and from Eq. (10.8) we have

$$\chi_{ijk}^{(2)}(-\omega_3, -\omega_1, -\omega_2) = \chi_{ijk}^{(2)}(\omega_3, \omega_1, \omega_2). \tag{10.10}$$

Second, in a lossless medium we can also interchange any two frequencies together with the corresponding indices:

$$\chi_{ijk}^{(2)}(\omega_3, \omega_1, \omega_2) = \chi_{jik}^{(2)}(-\omega_1, -\omega_3, \omega_2). \tag{10.11}$$

Importantly, the first frequency in the three arguments of the second-order susceptibility should be the sum of the other two. This property is called the *full permutation symmetry*.

But according to Eq. (10.10) the signs of all frequencies can be changed as well; therefore, we get

$$\chi_{ijk}^{(2)}(\omega_3, \omega_1, \omega_2) = \chi_{jik}^{(2)}(\omega_1, \omega_3, -\omega_2) = \chi_{kij}^{(2)}(\omega_2, \omega_3, -\omega_1). \tag{10.12}$$

Full permutation symmetry directly follows from the fact that the energy density of the electric field in a lossless medium is constant [2]. The proof is similar to the one we used in Chapter 4. Indeed, in the presence of nonlinear susceptibilities, expression (4.7) for the energy density of the electric field should be completed with nonlinear terms:

$$
\begin{aligned}
U_e &= \frac{\epsilon_0}{2} \sum_{ij} \sum_n \epsilon_{ij}(\omega_n) E_i^*(\omega_n) E_j(\omega_n) \\
&\quad + \frac{\epsilon_0}{3} \sum_{ijk} \sum_{nm} \chi_{ijk}^{(2)}(-\omega_n - \omega_m, \omega_n, \omega_m) E_i^*(\omega_n + \omega_m) E_j(\omega_n) E_k(\omega_m) \\
&\quad + \frac{\epsilon_0}{4} \sum_{ijkl} \sum_{nmo} \chi_{ijkl}^{(3)}(-\omega_n - \omega_m - \omega_o, \omega_n, \omega_m, \omega_o) \\
&\quad \times E_i^*(\omega_n + \omega_m + \omega_o) E_j(\omega_n) E_k(\omega_m) E_l(\omega_o) + \cdots.
\end{aligned} \tag{10.13}
$$

Because i, j, k, l are just dummy (summation) indices, the second-order and third-order susceptibilities $\chi_{ijk}^{(2)}(-\omega_n - \omega_m, \omega_n, \omega_m)$, $\chi_{ijkl}^{(3)}(-\omega_o - \omega_n - \omega_m, \omega_n, \omega_m, \omega_o)$ should have full permutation symmetry. Therefore, the susceptibilities with modified frequency arguments, $\chi_{ijk}^{(2)}(\omega_n + \omega_m, \omega_n, \omega_m)$ and $\chi_{ijkl}^{(3)}(\omega_o + \omega_n + \omega_m, \omega_n, \omega_m, \omega_o)$, should also have this property.

4. Kleinman's symmetry. The strongest symmetry restriction for the number of elements of the second-order susceptibility tensor is valid in materials without optical dispersion. The assumption that dispersion is absent, i. e., that the optical properties do not depend on the frequency, is even stronger than the assumption of no optical loss. This assumption is valid when any material resonances are far from the frequency range of interest. Typically, this is true in the middle of the visible–near-infrared range.

Then in Eq. (10.12) the sequence of frequencies is immaterial. One can therefore write

$$\chi_{ijk}^{(2)}(\omega_3, \omega_1, \omega_2) = \chi_{jik}^{(2)}(\omega_3, \omega_1, \omega_2) = \cdots = \chi_{kij}^{(2)}(\omega_3, \omega_1, \omega_2), \tag{10.14}$$

i. e., all possible permutations of the indices are allowed.

This assumption is really too strong and holds only approximately; but, for effects like second-harmonic generation, it is always possible to permute the last two indices. In this case, as well as in the presence of Kleinman's symmetry, contracted notation is used.

10.1.2 Contracted notation

Taking advantage of the intrinsic permutation symmetry, it is convenient to introduce *contracted notation*, where the sequence of the last two indices does not matter. In other words, we pass from the $\chi_{ijk}^{(2)}$ tensor to another tensor,

$$d_{il}(\omega_3, \omega_1, \omega_2) = \frac{1}{2}\chi_{ijk}^{(2)}(\omega_3, \omega_1, \omega_2), \tag{10.15}$$

where the factor $\frac{1}{2}$ is added for convenience and there is a correspondence between the combination of j, k indices and the l index:

$$\begin{array}{cccccccc} jk: & 11 & 22 & 33 & 23, 32 & 13, 31 & 12, 21 \\ l: & 1 & 2 & 3 & 4 & 5 & 6 \end{array} \tag{10.16}$$

Accordingly, the *nonlinear susceptibility tensor* has the form

$$d_{il} = \begin{pmatrix} d_{11} & d_{12} & d_{13} & d_{14} & d_{15} & d_{16} \\ d_{21} & d_{22} & d_{23} & d_{24} & d_{25} & d_{26} \\ d_{31} & d_{32} & d_{33} & d_{34} & d_{35} & d_{36} \end{pmatrix}. \tag{10.17}$$

If the Kleinman symmetry is valid, some of these 18 components turn out to be equal. For instance,

$$\begin{aligned} d_{12} &= \frac{1}{2}\chi_{122}^{(2)} = \frac{1}{2}\chi_{212}^{(2)} = d_{26}; \\ d_{13} &= \frac{1}{2}\chi_{133}^{(2)} = \frac{1}{2}\chi_{313}^{(2)} = d_{35}; \\ d_{23} &= \frac{1}{2}\chi_{233}^{(2)} = \frac{1}{2}\chi_{323}^{(2)} = d_{34}; \ldots. \end{aligned} \tag{10.18}$$

We find that in the presence of Kleinman's symmetry, there are only ten different components of the tensor:

$$d_{il} = \begin{pmatrix} d_{11} & d_{12} & d_{13} & d_{14} & d_{15} & d_{16} \\ d_{16} & d_{22} & d_{23} & d_{24} & d_{14} & d_{12} \\ d_{15} & d_{24} & d_{33} & d_{23} & d_{13} & d_{14} \end{pmatrix}. \tag{10.19}$$

Furthermore, for a certain crystal some components of this tensor will be zero according to the symmetry class to which the crystal belongs, in full agreement with the Neumann principle. The structure of the d_{il} tensor for different symmetry classes will be described in the next section.

10.1.3 Various crystal symmetries

In Section 4.2 we considered the linear optical properties for different crystal systems and classes. We saw that for crystals with the highest symmetry, the dielectric permittivity tensor has the simplest structure. The lower the crystal symmetry, the larger number of different elements of ϵ_{ij}. According to the Neumann principle, the situation with the second-order susceptibility is similar. We will now describe the properties of d_{il} for different types of crystals. So far, we are not assuming Kleinman's symmetry.

Isotropic crystals (cubic). One might think that, for cubic crystals, which have isotropic linear optical properties, there will be no second-order nonlinear effects. However, among the cubic crystals, only ones belonging to classes m$\bar{3}$ and m$\bar{3}$m are centrosymmetric and therefore have no second-order susceptibility. Crystals of another cubic class, 432, although being non-centrosymmetric, also have no second-order susceptibility because of other symmetry restrictions. But for crystals of classes 23 and $\bar{4}$3m (an example is gallium arsenide, GaAs), there are three nonzero elements of the d tensor: $d_{14} = d_{25} = d_{36}$. Moreover, the second-order susceptibility of GaAs is one of the highest known.

Uniaxial crystals. In the *hexagonal* system, classes 6/m and 6/mmm are centrosymmetric. For the rest, the d matrices have the structure

$$d_{il}^6 = \begin{pmatrix} 0 & 0 & 0 & d_{14} & d_{15} & 0 \\ 0 & 0 & 0 & d_{15} & -d_{14} & 0 \\ d_{31} & d_{31} & d_{33} & 0 & 0 & 0 \end{pmatrix} \tag{10.20}$$

for class 6 crystals;

$$d_{il}^{\bar{6}} = \begin{pmatrix} d_{11} & -d_{11} & 0 & 0 & 0 & d_{16} \\ d_{16} & -d_{16} & 0 & 0 & 0 & -d_{11} \\ 0 & 0 & 0 & 0 & 0 & 0 \end{pmatrix} \tag{10.21}$$

for class $\bar{6}$ crystals;

$$d_{il}^{622} = \begin{pmatrix} 0 & 0 & 0 & d_{14} & 0 & 0 \\ 0 & 0 & 0 & 0 & -d_{14} & 0 \\ 0 & 0 & 0 & 0 & 0 & 0 \end{pmatrix} \tag{10.22}$$

for class 622 crystals;

$$d_{il}^{6mm} = \begin{pmatrix} 0 & 0 & 0 & 0 & d_{15} & 0 \\ 0 & 0 & 0 & d_{15} & 0 & 0 \\ d_{31} & d_{31} & d_{33} & 0 & 0 & 0 \end{pmatrix} \tag{10.23}$$

for class 6mm crystals;

$$d_{il}^{\bar{6}m2} = \begin{pmatrix} 0 & 0 & 0 & 0 & 0 & d_{16} \\ d_{16} & -d_{16} & 0 & 0 & 0 & 0 \\ 0 & 0 & 0 & 0 & 0 & 0 \end{pmatrix} \tag{10.24}$$

for class $\bar{6}$m2 crystals.

Among *tetragonal* crystals, centrosymmetric are 4/m and 4/mmm ones. For the rest of the classes, the nonlinear tensors are

$$d_{il}^4 = d_{il}^6;$$

$$d_{il}^{\bar{4}} = \begin{pmatrix} 0 & 0 & 0 & d_{14} & d_{15} & 0 \\ 0 & 0 & 0 & -d_{15} & d_{14} & 0 \\ d_{31} & -d_{31} & 0 & 0 & 0 & d_{36} \end{pmatrix}; \tag{10.25}$$

$$d_{il}^{422} = d_{il}^{622};$$
$$d_{il}^{4mm} = d_{il}^{6mm};$$

$$d_{il}^{\bar{4}2m} = \begin{pmatrix} 0 & 0 & 0 & d_{14} & 0 & 0 \\ 0 & 0 & 0 & 0 & d_{14} & 0 \\ 0 & 0 & 0 & 0 & 0 & d_{36} \end{pmatrix}. \tag{10.26}$$

In the *trigonal system*, classes $\bar{3}$ and $\bar{3}$m are centrosymmetric and therefore have $d = 0$. For the other classes,

$$d_{il}^3 = \begin{pmatrix} d_{11} & -d_{11} & 0 & d_{14} & d_{15} & d_{16} \\ d_{16} & -d_{16} & 0 & d_{15} & -d_{14} & -d_{11} \\ d_{31} & d_{31} & d_{33} & 0 & 0 & 0 \end{pmatrix}; \tag{10.27}$$

$$d_{il}^{32} = \begin{pmatrix} d_{11} & -d_{11} & 0 & d_{14} & 0 & 0 \\ 0 & 0 & 0 & 0 & -d_{14} & -d_{11} \\ 0 & 0 & 0 & 0 & 0 & 0 \end{pmatrix}; \tag{10.28}$$

$$d_{il}^{3m} = \begin{pmatrix} 0 & 0 & 0 & 0 & d_{15} & d_{16} \\ d_{16} & -d_{16} & 0 & d_{15} & 0 & 0 \\ d_{31} & d_{31} & d_{33} & 0 & 0 & 0 \end{pmatrix}. \tag{10.29}$$

Biaxial crystals. In the *orthorhombic system*, crystals of class mmm are centrosymmetric. For the other classes,

$$d_{il}^{222} = \begin{pmatrix} 0 & 0 & 0 & d_{14} & 0 & 0 \\ 0 & 0 & 0 & 0 & d_{25} & 0 \\ 0 & 0 & 0 & 0 & 0 & d_{36} \end{pmatrix}; \tag{10.30}$$

$$d_{il}^{mm2} = \begin{pmatrix} 0 & 0 & 0 & 0 & d_{15} & 0 \\ 0 & 0 & 0 & d_{24} & 0 & 0 \\ d_{31} & d_{32} & d_{33} & 0 & 0 & 0 \end{pmatrix}. \tag{10.31}$$

Among *monoclinic* crystals, class 2/m is centrosymmetric. For the rest,

$$d_{il}^{2} = \begin{pmatrix} 0 & 0 & 0 & d_{14} & 0 & d_{16} \\ d_{21} & d_{22} & d_{23} & 0 & d_{25} & 0 \\ 0 & 0 & 0 & d_{34} & 0 & d_{36} \end{pmatrix}; \tag{10.32}$$

$$d_{il}^{m} = \begin{pmatrix} d_{11} & d_{12} & d_{13} & 0 & d_{15} & 0 \\ 0 & 0 & 0 & d_{24} & 0 & d_{26} \\ d_{31} & d_{32} & d_{33} & 0 & d_{35} & 0 \end{pmatrix}. \tag{10.33}$$

Finally, for *triclinic* crystals, class $\bar{1}$ is centrosymmetric, and for class 1 all elements of the d_{il} tensor are nonzero and, in the general case, different.

In the presence of Kleinman's symmetry, several components of d_{il} become zero or equal to each other [2].

10.1.4 Third-order susceptibility

Consider now third-order frequency conversion from frequencies ω_1, ω_2, and ω_3 to a certain frequency $\omega_4 \equiv \omega_1 + \omega_2 + \omega_3$. Similarly to Eq. (10.7), the third-order polarization at frequency ω_4 is

$$P_i^{(3)}(\omega_4) = \epsilon_0 \sum_{jkl} \chi_{ijkl}^{(3)}(\omega_4, \omega_1, \omega_2, \omega_3) E_j(\omega_1) E_k(\omega_2) E_l(\omega_3). \tag{10.34}$$

The intrinsic permutation symmetry will be valid here as well, i. e.,

$$\chi_{ijkl}^{(3)}(\omega_4, \omega_1, \omega_2, \omega_3) = \chi_{ikjl}^{(3)}(\omega_4, \omega_2, \omega_1, \omega_3) = \chi_{ijlk}^{(3)}(\omega_4, \omega_1, \omega_3, \omega_2) = \cdots \tag{10.35}$$

Also, similarly to the case of the second-order susceptibility, Kleinman's symmetry can be assumed if optical dispersion is negligible. This is again a very strong assumption, for instance, in the case of third-harmonic generation, where the frequencies involved differ three times.

Due to the intrinsic permutation symmetry, contracted notation can be also introduced for the cubic susceptibility. Instead of the rank 4 tensor $\chi_{ijkl}^{(3)}$, one can consider a 3×10 matrix c_{im} with $i = 1, 2, 3$ and $m = 1, \ldots, 0$:

$$c_{im}(\omega_4, \omega_1, \omega_2, \omega_3) = \chi_{ijkl}^{(3)}(\omega_4, \omega_1, \omega_2, \omega_3), \tag{10.36}$$

where the combination of j, k, l indices corresponds to the m index as [5]

$$
\begin{array}{ccccccccccc}
jkl: & 111 & 222 & 333 & 233, & 223, & 133, & 113, & 122, & 112, & 123, 132, \\
 & & & & 323, & 322, & 313, & 311, & 212, & 121, & 213, 231, \\
 & & & & 332 & 232 & 331 & 131 & 221 & 211 & 312, 321 \\
m: & 1 & 2 & 3 & 4 & 5 & 6 & 7 & 8 & 9 & 0
\end{array}
\tag{10.37}
$$

Similarly to the second-order susceptibility, the third-order susceptibility tensor has fewer independent elements in materials with higher symmetry. The simplest case is that of isotropic materials, for which $\chi_{ijkl}^{(3)}$ has only 21 nonzero element, only 3 of them independent. Namely, the nonzero values of the $\chi^{(3)}$ tensor are [2]

$$
\begin{aligned}
\chi_{xxxx}^{(3)} &= \chi_{yyyy}^{(3)} = \chi_{zzzz}^{(3)} \equiv \chi_1, \\
\chi_{xxyy}^{(3)} &= \chi_{yyxx}^{(3)} = \chi_{yyzz}^{(3)} = \chi_{zzyy}^{(3)} = \chi_{xxzz}^{(3)} = \chi_{zzxx}^{(3)} \equiv \chi_2, \\
\chi_{xyxy}^{(3)} &= \chi_{yxyx}^{(3)} = \chi_{yzyz}^{(3)} = \chi_{zyzy}^{(3)} = \chi_{xzxz}^{(3)} = \chi_{zxzx}^{(3)} \equiv \chi_3, \\
\chi_{xyyx}^{(3)} &= \chi_{yxxy}^{(3)} = \chi_{yzzy}^{(3)} = \chi_{zyyz}^{(3)} = \chi_{xzzx}^{(3)} = \chi_{zxxz}^{(3)} \equiv \chi_4,
\end{aligned}
\tag{10.38}
$$

related as

$$\chi_1 = \chi_2 + \chi_3 + \chi_4. \tag{10.39}$$

Equation (10.39) follows from the simple fact that in an anisotropic medium, the nonlinear effects are the same for any input polarization state. For instance, if the incident light is polarized diagonally and propagates along the z axis, $\vec{E} = E_0 \vec{e}_D$, then the third-order nonlinear polarization (for instance, for the third-harmonic generation) should be also polarized diagonally and its value should be

$$\vec{P}^{(3)} = \epsilon_0 \chi_1 E_0^3 \vec{e}_D. \tag{10.40}$$

At the same time, we can write it formally according to Eq. (10.34), assuming that the electric field has components $E_x = E_0/\sqrt{2}$ and $E_y = E_0/\sqrt{2}$. The components of the polarization vector $\vec{P}^{(3)}$, taking into account Eq. (10.38), will then be

$$
\begin{aligned}
P_x^{(3)} &= \epsilon_0 \frac{E_0^3}{2\sqrt{2}}(\chi_1 + \chi_2 + \chi_3 + \chi_4), \\
P_y^{(3)} &= \epsilon_0 \frac{E_0^3}{2\sqrt{2}}(\chi_1 + \chi_2 + \chi_3 + \chi_4).
\end{aligned}
\tag{10.41}
$$

From Eqs. (10.40), (10.41), we obtain Eq. (10.39).

The same way, i. e., by assuming that for an isotropic medium all polarization states of the input electric field are equivalent, we can derive an interesting feature: the third harmonic from a circularly polarized wave in an isotropic material is absent [6]. Indeed, let the incident field be right-hand circularly polarized, $\vec{E} = E_0 \vec{e}_R$. Then, as in the previous example, on the one hand,

$$\vec{P}^{(3)} = \epsilon_0 \chi_1 E_0^3 \vec{e}_R. \tag{10.42}$$

On the other hand, Eq. (10.34) yields in this case

$$P_x^{(3)} = \epsilon_0 \frac{E_0^3}{2\sqrt{2}} (\chi_1 - \chi_2 - \chi_3 - \chi_4),$$

$$P_y^{(3)} = -i\epsilon_0 \frac{E_0^3}{2\sqrt{2}} (\chi_1 - \chi_2 - \chi_3 - \chi_4). \tag{10.43}$$

From Eq. (10.39), it follows that $P_x^{(3)} = P_y^{(3)} = 0$. But even without Eq. (10.39), Eq. (10.43) is in contradiction with Eq. (10.42) because it describes a left-circularly polarized wave. Therefore, the third-harmonic generation should be absent.

Cubic crystals have the same 21 nonzero elements of the $\chi_{ijkl}^{(3)}$ tensor as isotropic materials, but more of them independent: 4 for some classes and 7 for others [2]. For instance, for classes 432, $\bar{4}3m$, and $m3m$, there are the same 4 nonzero elements (10.38) as for isotropic materials, but without Eq. (10.39). Further, the less symmetric a crystal is, the more nonzero and independent elements the tensor $\chi^{(3)}$ has. The least symmetric, triclinic crystals have 81 nonzero elements of the $\chi^{(3)}$ tensor, all of them independent. A detailed table of nonzero elements of $\chi_{ijkl}^{(3)}$ for each crystal group can be found in book [2].

10.2 Phase matching

10.2.1 Helmholtz equation

In the general case of a nonlinear interaction, all participating waves can have different polarization states. For instance, in the simplest case of the second-harmonic generation, Eq. (10.7) becomes

$$P_i^{(2)}(2\omega) = \epsilon_0 \sum_{jk} \chi_{ijk}^{(2)}(2\omega, \omega, \omega) E_j(\omega) E_k(\omega), \tag{10.44}$$

or, using the nonlinear susceptibility tensor,

$$\begin{pmatrix} P_x^{(2)}(2\omega) \\ P_y^{(2)}(2\omega) \\ P_z^{(2)}(2\omega) \end{pmatrix} = 2\epsilon_0 \begin{pmatrix} d_{11} & d_{12} & d_{13} & d_{14} & d_{15} & d_{16} \\ d_{16} & d_{22} & d_{23} & d_{24} & d_{14} & d_{12} \\ d_{15} & d_{24} & d_{33} & d_{23} & d_{13} & d_{14} \end{pmatrix} \begin{pmatrix} E_x^2(\omega) \\ E_y^2(\omega) \\ E_z^2(\omega) \\ 2E_y(\omega)E_z(\omega) \\ 2E_x(\omega)E_z(\omega) \\ 2E_x(\omega)E_y(\omega) \end{pmatrix}. \tag{10.45}$$

It follows that the direction of the nonlinear polarization vector $\vec{P}^{(2)}(2\omega)$ depends on both the structure of the nonlinear tensor d_{il} and the polarization of the fundamental harmonic wave. The same will be valid for any second-order or third-order nonlinear process.

How will the electromagnetic waves emerging due to the nonlinear interaction be polarized? For instance, what will be the polarization state of the second harmonic resulting from nonlinear polarization (10.44)? The approach used in nonlinear optics is based on the Helmholtz equation describing the electric field wave $\vec{E}$ induced by the nonlinear polarization wave:

$$\nabla^2 \vec{E} - \frac{\epsilon}{c^2} \frac{\partial^2 \vec{E}}{\partial t^2} = \frac{1}{\epsilon_0 c^2} \frac{\partial^2 \vec{P}^{NL}}{\partial t^2}. \tag{10.46}$$

This Helmholtz equation differs from the one we derived in Chapter 7 by the right-hand side: now it is nonzero. This is a typical equation describing oscillations of a system (electric field) due to the driving force (nonlinear polarization). At each point, the electric field will have the same direction and the same oscillation frequency as the polarization (10.45). At the same time, the propagation of a wave $\vec{E}(\vec{r}, t)$ in a material requires certain conditions to be satisfied. First, the wavevector and the frequency of a propagating wave should obey the dispersion relation $\vec{k} = \vec{k}(\omega)$. Second, as it was shown in Chapter 4, only waves in two polarization states can propagate in a crystal: the ordinary wave and the extraordinary wave. The first restriction leads to the *phase matching condition*. The second one dictates the possible polarization *types of phase matching*.

The wave vectors of the nonlinear polarization wave and the induced field wave are, in the general case, different. For instance, in the case of second-harmonic generation, the incident field at frequency ω with the wavevector $\vec{k}(\omega)$ will generate a wave of second-order nonlinear polarization $\vec{P}^{(2)}$ at frequency 2ω. This polarization wave will propagate with the wavevector $2\vec{k}(\omega)$, but the induced electric field wave at frequency 2ω can only propagate with a certain wavevector $\vec{k}(2\omega)$, satisfying the dispersion relation in the medium. In the presence of dispersion, the refractive index depends on the frequency, and the wave vector is (comp. with Eq. (4.14), which did not take dispersion into account)

$$k(\omega) = \frac{n(\omega)\omega}{c}. \tag{10.47}$$

Therefore, in the presence of dispersion the nonlinear polarization and the electric field will have different wavevectors, whose absolute values are $2k(\omega) = 2\frac{n(\omega)\omega}{c}$ and $k(2\omega) = 2\frac{n(2\omega)\omega}{c}$, respectively. Unless $n(\omega) = n(2\omega)$, the waves of the electric field $\vec{E}(\vec{r},t)$ and the nonlinear polarization $\vec{P}^{(2)}(\vec{r},t)$ will propagate with different phase velocities and get out of phase at some point. Then the interaction becomes inefficient. This explains, in simple terms, why nonlinear optical processes require phase matching. In particular, for second-harmonic generation the phase matching requires the condition

$$n(2\omega) = n(\omega), \tag{10.48}$$

and for third-harmonic generation, condition

$$n(3\omega) = n(\omega). \tag{10.49}$$

For sum- and difference-frequency generation, the conditions are more complicated:

$$\omega_1[n(\omega_1 \pm \omega_2) - n(\omega_1)] = \pm\omega_2[n(\omega_2) - n(\omega_1 \pm \omega_2)]. \tag{10.50}$$

10.2.2 Types of phase matching

Each of conditions (10.48), (10.49), (10.50) are impossible to satisfy under normal dispersion where $n(\omega)$ increases with ω. (For instance, in Eq. (10.50), the left-hand side and right-hand side are of opposite signs.) As an example, Fig. 10.1 shows the dispersion of ordinary and extraordinary refractive indices in lithium niobate crystal. If the second harmonic has to be generated from the radiation at $1.064\,\mu$ (the wavelength of a Nd:YAG laser), the refractive indices for the second-harmonic and fundamental radiation will differ by almost 0.1, and the phase matching condition (10.48) will not be satisfied.

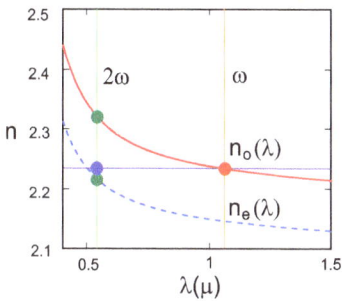

Figure 10.1: Dispersion of ordinary (red solid line) and extraordinary (blue dotted line) refractive indices in lithium niobate crystal and the effective refractive index (blue dot) for the second-harmonic radiation at $0.532\,\mu$ (2ω) generated under type-I phase matching.

However, the phase matching conditions for second-harmonic, sum- and difference-frequency generation, and sometimes even for third-harmonic generation can be satisfied by using crystal birefringence and requiring that the fields at different frequencies should be polarized differently.

Consider, for instance, second-harmonic generation in a uniaxial crystal. Let the fundamental wave at frequency ω be incident at some angle ϑ to the optic axis z (Fig. 10.2). It is further convenient to consider only the plane formed by the wavevector $\vec{k}$ and the optic axis z. From Chapter 4 (see Fig. 4.3), we know that only two polarization states are allowed for any $\vec{k}$ direction: the ordinary wave is polarized orthogonally to the $(\vec{k}, z)$ plane (green vector) and the extraordinary wave is polarized parallel to this plane (blue vector). The wavevector of the second-harmonic radiation is collinear to $\vec{k}$; it follows that, for the second-harmonic wave, the same two polarization states are possible.

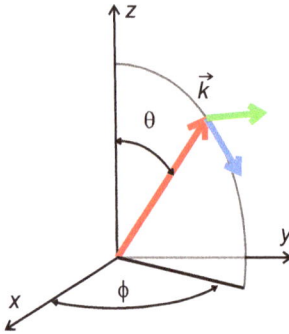

Figure 10.2: Orientation of the wave vector (red) and the polarization directions (green, blue) for second-harmonic generation in a uniaxial crystal.

Depending on the polarization states of the fundamental wave at frequency ω and the second-harmonic wave at frequency 2ω, different types of interaction are distinguished:

(i) oo→e interaction: the fundamental wave is ordinary polarized and the second harmonic is extraordinary polarized.

(ii) ee→o interaction: the fundamental wave is extraordinary polarized and the second harmonic is ordinary polarized.

(iii) ee→e interaction: both waves are extraordinary polarized.

(iv) oo→o interaction: both waves are ordinary polarized.

(v) eo→e interaction: one fundamental-radiation photon is ordinary polarized, the other one is extraordinary polarized, the second harmonic is extraordinary polarized.

(vi) eo→o interaction: the same as (v) but the second harmonic is ordinary polarized.

The cases (i) and (ii) are called type-I interaction, the cases (v) and (vi), type-II interaction, and the cases (iii) and (iv), type-0 interactions.

Some of these interactions can help to satisfy condition (10.48). For instance, for second-harmonic generation from $1.064\,\mu$ to $0.532\,\mu$ in lithium niobate (Fig. 10.1), the dispersion in the visible range is somewhat smaller than birefringence. We see that, as it should be for normal dispersion, $n_o(\omega) < n_o(2\omega)$ and $n_e(\omega) < n_e(2\omega)$, so type-0 interaction is impossible. However (see the two green points and the red point in Fig. 10.1), because lithium niobate is a negative crystal, $n_e(2\omega) < n_o(\omega) < n_o(2\omega)$. By properly choosing the angle ϑ to the optic axis one can make the effective index [see Eq. (4.43) of Chapter 4],

$$n(2\omega) = \left(\frac{\sin^2 \vartheta}{n_e^2(2\omega)} + \frac{\cos^2 \vartheta}{n_o^2(2\omega)} \right)^{-1/2}, \tag{10.51}$$

take any value between $n_e(2\omega)$ and $n_o(2\omega)$.

Then, one can make $n(2\omega) = n_o(\omega)$ (blue point in Fig. 10.1), and the phase matching is satisfied for type-I interaction oo→e. From Eq. (10.51) we find that the angle between the wave vector and the optic axis should be

$$\vartheta = \arcsin\left(\sqrt{\frac{[n_o(2\omega)/n_o(\omega)]^2 - 1}{[n_o(2\omega)/n_e(2\omega)]^2 - 1}} \right). \tag{10.52}$$

Using the values of refractive indices for lithium niobate doped with magnesium[2] [8] $n_o(0.532\,\mu) = 2.3232$, $n_e(0.532\,\mu) = 2.2336$, and $n_o(1.064\,\mu) = 2.2321$, we find that, for phase-matched oo→e second-harmonic generation, $\vartheta = 82.6\,\mathrm{deg}$.

For a positive uniaxial crystal, this strategy will not work because the effective refractive index $n(2\omega)$ will be always larger than $n_o(\omega)$. However, the type I phase matching ee→o can be satisfied in this case: the incident wave has to be extraordinary polarized, and the second harmonic, ordinary polarized.

For some third-order nonlinear interactions, phase matching can be also satisfied using birefringence. Even for third-harmonic generation, where the dispersion leads to a huge difference between the refractive indices of the interacting waves, $n(\omega)$ and $n(3\omega)$, large birefringence of some crystals can suffice to satisfy the phase matching. The following types of phase matching are distinguished in this case [5]:

Type I: eee→o (in positive uniaxial crystals) or ooo→e (in negative uniaxial crystals);

Type II: eeo→o (in positive uniaxial crystals) or ooe→e (in negative uniaxial crystals);

Type III: eoo→o (in positive uniaxial crystals) or oee→e (in negative uniaxial crystals).

For instance, type-II ooe→e third-harmonic generation can be observed in calcite [7]. In titanium dioxide (rutile) crystal, both type-II eeo→o and type-III eee→o phase matching are possible [4]. In practice, however, the radiation frequency is usually tripled not through direct third-harmonic radiation but through a cascade of second-order processes: first, second-harmonic generation $\omega \rightarrow 2\omega$ and then sum-frequency generation, $\omega, 2\omega \rightarrow 3\omega$.

---

[2] Doping lithium niobate with magnesium (typically, 5 %) reduces the photorefractive effect and is therefore widely used in nonlinear optics.

10.2.3 Effective susceptibility

Now, with the geometry of a nonlinear optical process dictated by the phase matching, and the nonlinear susceptibility tensor components determined by the symmetry of the nonlinear material, we can find the effective value of the nonlinear susceptibility. This can be done with the help of Eq. (10.45) [3]. In the example considered above, lithium niobate is a class 3m crystal, and its d tensor has the form (10.29). Then Eq. (10.45) takes the form

$$
\begin{pmatrix} P_x^{(2)}(2\omega) \\ P_y^{(2)}(2\omega) \\ P_z^{(2)}(2\omega) \end{pmatrix} = 2\epsilon_0 \begin{pmatrix} 0 & 0 & 0 & 0 & d_{15} & d_{16} \\ d_{16} & -d_{16} & 0 & d_{15} & 0 & 0 \\ d_{15} & d_{15} & d_{33} & 0 & 0 & 0 \end{pmatrix} \begin{pmatrix} E_x^2(\omega) \\ E_y^2(\omega) \\ E_z^2(\omega) \\ 2E_y(\omega)E_z(\omega) \\ 2E_x(\omega)E_z(\omega) \\ 2E_x(\omega)E_y(\omega) \end{pmatrix}. \quad (10.53)
$$

If oo→e geometry is used, the fundamental wave is ordinary polarized, which means that its electric field vector has the components

$$
\vec{E}(\omega) = E(\omega) \begin{pmatrix} -\sin\varphi \\ \cos\varphi \\ 0 \end{pmatrix}, \quad (10.54)
$$

where φ is the azimuthal angle of the incident wave (see Fig. 10.2).
The components of the nonlinear polarization (10.53) are then

$$
\begin{aligned}
P_x(2\omega) &= -2\epsilon_0 d_{16} E^2(\omega)\sin(2\varphi), \\
P_y(2\omega) &= -2\epsilon_0 d_{16} E^2(\omega)\cos(2\varphi), \\
P_z(2\omega) &= 2\epsilon_0 d_{15} E^2(\omega).
\end{aligned} \quad (10.55)
$$

The second-harmonic radiation has to be extraordinary polarized. This means that its electric field vector should be along the unit vector with the components

$$
\vec{e} \equiv \begin{pmatrix} \cos\vartheta\cos\varphi \\ \cos\vartheta\sin\varphi \\ -\sin\vartheta \end{pmatrix}, \quad (10.56)
$$

where we have ignored the spatial walk-off and assumed that the electric field of the extraordinary wave is orthogonal to the wavevector. The extraodinary polarized electric field will be determined by the projection of the nonlinear polarization at frequency 2ω (10.55) on the unit vector (10.56),

$$
P(2\omega)_{\text{eff}} = -2\epsilon_0[d_{16}\sin(3\varphi)\cos\vartheta + d_{15}\sin\vartheta]E^2(\omega). \quad (10.57)
$$

Then the effective nonlinearity for the oo→e phase matching is

$$d_{\text{eff}} = -[d_{16} \sin(3\varphi) \cos \vartheta + d_{15} \sin \vartheta]. \tag{10.58}$$

Similarly, one can find effective nonlinearities for any type of nonlinear interaction. Expressions for d_{eff} for all symmetry classes and for all types of interaction can be found in Refs. [3, 9].

This result can be generalized by recalling the whole procedure: we find the nonlinear polarization vector by multiplying the $\hat{\chi}^{(2)}$ tensor by two vectors corresponding to the pump polarization, $\vec{P} \propto \hat{\chi}^{(2)} : \vec{E}\,\vec{E}$. Then we project this vector on the unit vector corresponding to the polarization direction of the ordinary or extraordinary wave in the crystal: $\vec{P} \cdot \vec{e}$. As a result, the effective second-order susceptibility is

$$\chi_{\text{eff}}^{(2)} = \hat{\chi}^{(2)} : \vec{e}_{2\omega}\vec{e}_{\omega}\vec{e}_{\omega}, \tag{10.59}$$

where $\vec{e}_{2\omega}$ and $\vec{e}_{\omega}$ are unit vectors corresponding to the polarization directions (ordinary or extraordinary) of the incident and second-harmonic waves, and the three dots, as before, denote the multiplication of a tensor by three vectors. Then the effective nonlinearity is related to the second-order susceptibility as $d_{\text{eff}} = \frac{1}{2}\chi_{\text{eff}}^{(2)}$.

Generalizing Eq. (10.59) to the case of an arbitrary three-wave interaction, we obtain

$$d_{\text{eff}} = \frac{1}{2}\hat{\chi}^{(2)} : \vec{e}_1\vec{e}_2\vec{e}_3, \tag{10.60}$$

where the indices label the three interacting waves.

It is important to note that the effective nonlinearity can be positive or negative, depending on the direction of the electric field vector with respect to the crystal axes. Of course the electric field in an electromagnetic wave oscillates, and so do the signs of its projections on the crystal axes. If the nonlinear interaction occurs in just one crystal, the sign of d_{eff} does not matter. But if there are two or more nonlinear crystals interacting with the same incident wave, their mutual orientation matters. Depending on whether their effective nonlinearities have the same or opposite signs, nonlinear processes in two neighboring crystals can enhance or cancel each other. This effect is called *nonlinear interference*.

In particular, in a *periodically poled crystal* the *spontaneous polarization* (z) axis changes its direction in the neighboring areas (*domains*). As a result, the sign of the second-order nonlinearity is changed with a certain period. This variation can be used to satisfy the type-0 phase matching [2], which enables access to the largest d_{33} component of the nonlinear tensor in crystals like lithium niobate, lithium tantalate, and potassium titanyl phosphate (KTP). This method is called *quasi-phasematching* [2]; see Section 10.3.3.

10.3 The effect of spatial walk-off and its elimination

In the presence of spatial walk-off (Section 4.3), different beams can get not only out of phase, but also spatially displaced from each other in the course of nonlinear interaction if they are orthogonally polarized. In the cases of second-harmonic, sum- and difference-frequency generation, and parametric amplification, this will happen for both type-I and type-II interactions. This effect becomes especially pronounced under relatively tight focusing, typical in nonlinear optics experiments, and does not allow one to use long nonlinear crystals in this case. In order to provide efficient interaction, the length L of a nonlinear crystal should not exceed the value $a/\tan\alpha$, where a is the diameter of the beam and α the walk-off angle. Fortunately, there are several methods to avoid spatial walk-off.

10.3.1 Walk-off compensation

One way to reduce the detrimental effect of spatial walk-off is to use, instead of a long crystal, two crystals with optic axes tilted symmetrically. Figure 10.3 shows this arrangement for the case of type-II second-harmonic generation [10]. To provide type-II interaction, the fundamental wave (ω) should be polarized at 45° to the plane formed by the optic axis (z) and the incident wave vector direction. Inside the crystal the incident beam splits in two, the ordinary beam (o) propagating in the original direction, and the extraordinary beam (e) tilted by the walk-off angle. If a single long crystal is used (panel a), the two beams are strongly displaced from each other in the end of the crystal, and the second harmonic (2ω) is generated only at the very beginning. Instead, one can place two shorter crystals after one another (panel b), so that their optic axes are both oriented at the phase matching angle ϑ to the incident wavevector, but tilted symmetrically. Then the walk-off directions in the two crystals will be opposite, and the shift of the extraordinary wave in the first crystal will be compensated for in the second crystal. As a result, the range of efficient second-harmonic generation will be increased by a factor of two.

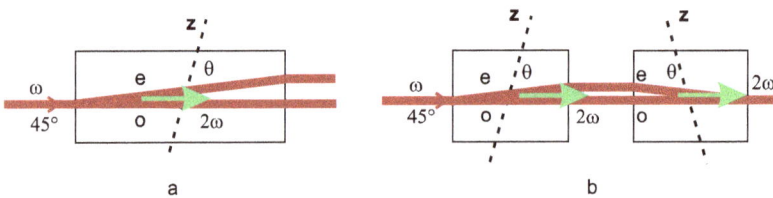

Figure 10.3: Spatial walk-off in type-II second-harmonic generation (a) and its compensation by using two shorter crystals with symmetric orientations of the optic axes (b).

This scheme of walk-off compensation has been implemented for second-harmonic generation in KTP crystals in Ref. [10]. Somewhat later, the same principle has been applied to parametric amplification and oscillation [1].

10.3.2 Non-critical phase matching

As shown in Section 4.3, the amount of spatial walk-off depends on the angle ϑ between the wavevector and the optic axis. It is maximal if ϑ is 45° but it is absent if ϑ is 0° or 90°. Accordingly, a way to completely eliminate the walk-off effect is to find phase matching at $\vartheta = 90°$. For instance, type-I phase matching for second-harmonic generation from 1.064 μ in lithium niobate (Section 10.2) occurs at an angle $\vartheta = 82.6°$, i. e., in the direction almost orthogonal to the optic axis. It turns out that by changing the temperature of the crystal, this angle can be made exactly 90° [2]. This type of phase matching, called *non-critical phase matching*, is possible because the ordinary and extraordinary refractive indices of lithium niobate depend on the temperature differently. Temperature-tunable non-critical phase matching is often used in lithium niobate, KTP, lithium triborate (LBO) and several other crystals. Unlike the compensation method considered above, it enables complete elimination of the walk-off effect.

10.3.3 Quasi-phasematching

In Section 10.2.3 we already mentioned quasi-phasematching in periodically poled crystals. It is worth mentioning here that this technique also eliminates the spatial walk-off. Indeed, in a periodically poled crystal the direction of the optic axis is flipped in the neighboring domains; at the same time, the optic axis is always parallel to the domain walls. Light propagates orthogonally to the domain walls and hence orthogonally to the optic axis direction. The spatial walk-off therefore does not occur. This situation is shown in Fig. 10.4. One can say that in each domain the phase matching is non-critical; however, instead of temperature tuning, described in the previous section, here the phase matching is tuned by varying the poling period [2].

Quasi-phasematched nonlinear interaction will be efficient even in long crystals. For instance, lithium niobate crystals used in nonlinear optics can be as long as a few cm.

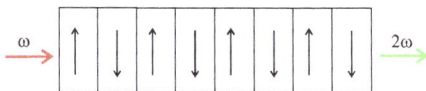

Figure 10.4: Second-harmonic generation in a periodically poled crystal. Black arrows show the optic axis directions in different domains.

Bibliography

[1]   D. J. Armstrong, W. J. Alford, T. D. Raymond, A. V. Smith, and M. S. Bowers. Parametric amplification and oscillation with walkoff-compensating crystals. *J. Opt. Soc. Am. B*, 1997.

[2]   R. W. Boyd. *Nonlinear optics*. Academic Press, 2008.

[3]   V. G. Dmitriev, G. G. Gurzadyan, and D. N. Nikogosyan. *Handbook of nonlinear crystals*. Springer, 1999.

[4]   F. Gravier and B. Boulanger. Cubic parametric frequency generation in rutile single crystal. *Opt. Express*, 14(24):11715–11720, Nov 2006.

[5]   J. E. Midwinter and J. Warner. The effects of phase matching method and of crystal symmetry on the polar dependence of third order non-linear optical polarization. *Br. J. Appl. Phys.*, 16:1667–1674, 1965.

[6]   G. New. *Introduction to nonlinear optics*. Cambridge University Press, 2011.

[7]   A. Penzkofer, F. Ossig, and P. Qiu. Picosecond third-harmonic light generation in calcite. *Appl. Phys. B*, 47:71–81, 1988.

[8]   D. E. Zelmon, D. L. Small, and D. Jundt. Infrared corrected Sellmeier coefficients for congruently grown lithium niobate and 5 mol. % magnesium oxide-doped lithium niobate. *J. Opt. Soc. Am. B*, 14(12):3319–3322, Dec 1997.

[9]   F. Zernike and J. E. Midwinter. *Applied nonlinear optics*. John Wiley and Sons, 1973.

[10] J.-J. Zondy, M. Abed, and S. Khodja. Twin-crystal walk-off-compensated type-II second-harmonic generation: single-pass and cavity-enhanced experiments in ktiopo4. *J. Opt. Soc. Am. B*, 11(12):2368–2379, Dec 1994.

11 Quantum description of polarization

In this chapter we will re-consider the description of polarization used so far throughout the book. In quantum optics, like generally in quantum mechanics, every physical observable corresponds to some *operator*. Accordingly, in this chapter, instead of the Stokes observables we will introduce the Stokes operators. Unlike classical Stokes observables, they will be defined not in terms of intensities, but in terms of *photon-number operators*. Instead of electric fields, which were the basic classical observables of the previous chapters, we will now use *electric field operators*, and further, *photon-creation and -annihilation operators*. All these operators will act on the quantum states of polarized light. In this chapter, we will mainly consider single-photon polarized states; more complicated states of polarized light will be the subject of Chapter 12.

The reader of this chapter is expected to have studied quantum mechanics, but not necessarily quantum optics. Therefore, the basic notions and instruments of quantum optics will be briefly introduced here from the polarization point of view.

11.1 Basic notions of quantum optics

Quantum optics typically deals with quantum states of light defined in one or two modes. A single mode is specified by a direction (plane-wave mode) and a wavelength (monochromatic mode). Polarization adds to this picture another degree of freedom because for any wavelength and for any direction, there are also two orthogonal polarization states. Throughout Chapters 11 and 12, we will consider one or two plane-wave monochromatic modes and two polarization modes. In the following subsection, we will briefly review how monochromatic plane-wave modes are introduced.

11.1.1 Modes of electromagnetic field and field quantization

Modes of electromagnetic fields are introduced as domains in real space, or in the reciprocal space given by wavevector $\vec{k}$, where the field is coherent with itself but incoherent with the field in other domains. The simplest way to introduce modes is as plane monochromatic waves; however, one can choose other ways: spherical waves, Gaussian beams, or broadband coherent modes [23]. In each case, the set of modes is formed by orthogonal functions in real or reciprocal space. Here we will consider the simplest case of plane-wave monochromatic modes. To introduce them, we need to choose a *quantization box*—this is simply a very large box in space such that 'nothing interesting' happens outside of it. In particular, we can always choose the box so large that all electromagnetic fields outside this box are zero. Then it makes no difference to assume that the distribution of fields inside the box is periodically repeated outside

https://doi.org/10.1515/9783110668025-011

it [15]. By assuming that the field distribution is periodic in all three Cartesian coordinates x, y, z, with the period given by the box size L, we obtain the condition that only a discrete set of wavevectors are allowed:

$$\vec{k} \equiv \{k_x; k_y; k_z\} = \frac{2\pi}{L}\{l; m; n\}, \tag{11.1}$$

where l, m, n are integer numbers.

These discrete modes are shown in Fig. 11.1. To specify such a mode, we need to fix three components of the wavevector. Alternatively, we can specify the direction of the wavevector and its absolute value, related to the frequency ω via the dispersion dependence, $k = n(\omega)\omega/c$.

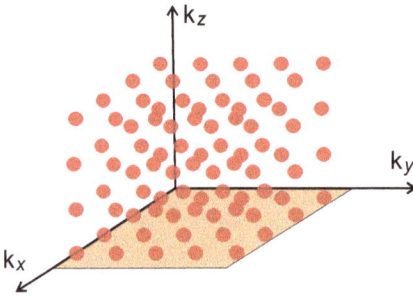

Figure 11.1: Plane-wave monochromatic modes in the wavevector space.

But according to Chapter 3, for every such a mode, i. e., for every direction and absolute value of the wavevector, there are also two polarization states. Then for completeness, one should specify a mode by four, rather than three numbers,

$$\vec{k} = \{k_x; k_y, k_z, \sigma\}, \tag{11.2}$$

where σ denotes the polarization and can take two values. Depending on the situation, the two polarization modes can be linear, or circular, or elliptic, but they should be orthogonal. Which polarization basis to use depends on the specific problem: for instance, in a crystal, it is more convenient to use linearly polarized modes corresponding to the ordinary and extraordinary beams. In an optically active material, the convenient polarization mode basis will comprise left- and right-polarized modes, L and R. But in the laboratory frame of reference, most convenient are horizontally and vertically polarized modes. In what follows, we will usually consider one wavevector mode and two polarization modes: H and V, or D and A, or L and R.

The complex fields in classical optics, the negative-frequency one $E^{(-)}$ [Eq. (2.2)] and the positive-frequency one, $E^{(+)}$ [Eq. (2.3)], in quantum optics become operators, $\hat{E}^{(-)}$ and $\hat{E}^{(+)}$, respectively. These operators are, in turn, written as superpositions over modes,

$$\hat{E}^{(+)}(\vec{r}, t) = \sum_{\vec{k}} \hat{E}^{(+)}_{\vec{k}} e^{i\vec{k}\vec{r} - i\omega_{\vec{k}} t}, \tag{11.3}$$

where $\vec{k}$ is given by Eq. (11.2), the amplitude $\hat{E}_{\vec{k}}^{(+)} = c_{\vec{k}} a_{\vec{k}}$, $a_{\vec{k}}$ is the *photon-annihilation operator* in mode $\vec{k}$, $\omega_{\vec{k}}$ the frequency of this mode, and $c_{\vec{k}} = i\sqrt{\frac{\hbar\omega_{\vec{k}}}{2\epsilon_0 L^3}}$ in free space [23]. Summation over wavevectors $\vec{k}$ implies summation over the indices l, m, n [see (11.1)] and two polarization states. Similarly, the negative-frequency field operator is Hermitian conjugated to the positive-frequency one. We obtain

$$\hat{E}^{(+)}(\vec{r},t) = \sum_{\vec{k}} c_{\vec{k}} a_{\vec{k}} e^{i\vec{k}\vec{r}-i\omega_{\vec{k}}t},$$

$$\hat{E}^{(-)}(\vec{r},t) = \sum_{\vec{k}} c_{\vec{k}}^* a_{\vec{k}}^\dagger e^{-i\vec{k}\vec{r}+i\omega_{\vec{k}}t}, \tag{11.4}$$

where $a_{\vec{k}}^\dagger$ is the *photon-creation operator* in mode $\vec{k}$. Being dimensionless, photon-creation and -annihilation operators are more convenient than electric field operators.

It follows from the Maxwell equations [23] that each mode of the electromagnetic field is similar to a harmonic oscillator and, as such, can be populated by a certain number of quanta (photons). The energy operator, or *Hamiltonian*, of each mode can be written as

$$\hat{\mathcal{H}}_{\vec{k}} = \hbar\omega_{\vec{k}}\left(a_{\vec{k}}^\dagger a_{\vec{k}} + \frac{1}{2}\right). \tag{11.5}$$

11.1.2 Operators and their eigenstates

Throughout this chapter and Chapter 12, the focus of our discussion will be on the polarization modes, usually H and V. Accordingly, there will be two photon-annihilation operators, $a_{H,V}$ and their Hermitian conjugates $a_{H,V}^\dagger$.

These operators are not Hermitian: $a_{H,V}^\dagger \neq a_{H,V}$. Photon creation and annihilation operators for each mode do not commute:

$$[a_H, a_H^\dagger] = [a_V, a_V^\dagger] = 1. \tag{11.6}$$

At the same time, the operators for different modes do commute:

$$[a_H, a_V^\dagger] = [a_H, a_V] = \cdots = 0. \tag{11.7}$$

This is because polarization modes are orthogonal.

Although quantum field operators differ a lot from their classical counterparts, they still have many similar properties. In particular, passing from one polarization basis to another is performed via the same transformations as in classical polarization optics. For instance, photon-annihilation operators for the diagonal/antidiagonal and right- and left-circular polarization modes can be written as

$$a_{D,A} = \frac{a_H \pm a_V}{\sqrt{2}},$$

$$a_{R,L} = \frac{a_H \pm i a_V}{\sqrt{2}}, \tag{11.8}$$

[compare with Eqs. (3.9), (3.10)] and the corresponding expressions for photon-creation operators are obtained by Hermitian conjugation.

Being non-Hermitian, photon-creation and -annihilation operators do not correspond to observable quantities. This is not surprising because their classical counterparts, negative- and positive-frequency fields, are complex and therefore cannot be measured. To obtain Hermitian operators from $a_{H,V}^\dagger$, $a_{H,V}$, one can take their Hermitian and anti-Hermitian parts and construct *quadrature operators* for both polarization modes:

$$\hat{q}_{H,V} \equiv \frac{1}{2}(a_{H,V} + a_{H,V}^\dagger), \quad \hat{p}_{H,V} \equiv \frac{1}{2i}(a_{H,V} - a_{H,V}^\dagger). \tag{11.9}$$

Classical counterparts of the quadrature operators are cosine and sine quadratures of the electric field.

It is useful to introduce a generalized quadrature. For each mode, it has the form

$$\hat{q}_\varphi \equiv \hat{q} \cos \varphi + \hat{p} \sin \varphi, \tag{11.10}$$

where $0 \le \varphi < \pi$.

Photon creation and annihilation operators form the photon-number operators for each mode:

$$a_H^\dagger a_H \equiv \hat{N}_H, \quad a_V^\dagger a_V \equiv \hat{N}_V. \tag{11.11}$$

The photon-number operators are Hermitian,

$$\hat{N}_H^\dagger = \hat{N}_H, \quad \hat{N}_V^\dagger = \hat{N}_V. \tag{11.12}$$

As any Hermitian operators, they correspond to real observables, which can be measured. Indeed, according to Eq. (2.6), the classical counterparts of photon-number operators $\hat{N}_{H,V}$ are intensities in the polarization modes.

Quantum states are often introduced in quantum mechanics as eigenstates of various operators. Here we will briefly describe the states further used in this book. Note that because the two polarization modes are orthogonal, in the simplest case a state $|\Psi\rangle$ is a direct product of states in the two polarization modes: $|\Psi\rangle = |\Psi_H\rangle_H \otimes |\Psi_V\rangle_V$, or simply $|\Psi\rangle = |\Psi_H\rangle_H |\Psi_V\rangle_V$.

Fock states. The eigenvalues of photon-number operators $\hat{N}_{H,V}$ are non-negative integer numbers $N_{H,V}$, and their eigenstates are so-called Fock states, or number states:

$$\hat{N}_{H,V} |N\rangle_{H,V} = N_{H,V} |N\rangle_{H,V}. \tag{11.13}$$

(We denote, as is common in quantum mechanics, the operator, its eigenvalue, and its eigenstate by the same character.) Fock states are states in which the number of photons populating a mode is fixed and does not fluctuate. In practice, among all Fock states, only single-photon and two-photon ones can be prepared in laboratories in a relatively simple way.

Fock states of each mode form a complete orthonormal basis, and any state can be decomposed over this basis. The possibility of such a decomposition leads to the *decomposition of the identity* operator:

$$\hat{1} = \sum_N |N\rangle\langle N|, \tag{11.14}$$

which is valid for the Fock states in each polarization mode.

It is useful to recall here how the photon-creation and -annihilation operators act on Fock states:

$$a_{H,V}|N\rangle_{H,V} = \sqrt{N}|N-1\rangle_{H,V}, \quad a^{\dagger}_{H,V}|N\rangle_{H,V} = \sqrt{N+1}|N+1\rangle_{H,V}. \tag{11.15}$$

These so-called *ladder equations* describe the transitions between different states of a harmonic oscillator populated by a fixed number of quanta.

Coherent states are defined as the eigenstates of photon-annihilation operators $a_{H,V}$:

$$a_{H,V}|\alpha\rangle_{H,V} = \alpha|\alpha\rangle_{H,V}. \tag{11.16}$$

Similarly,

$$\langle\alpha|_{H,V} a^{\dagger}_{H,V} = \alpha^* \langle\alpha|_{H,V}. \tag{11.17}$$

The eigenvalues of coherent states are complex. Being eigenstates of a non-Hermitian operator, coherent states do not form an orthonormal set. Indeed, the inner product of two coherent states is nonzero [21],

$$\left|\langle\alpha|\beta\rangle\right|^2 = e^{-|\alpha-\beta|^2}. \tag{11.18}$$

Coherent states form a convenient basis, known as the Glauber–Sudarshan representation [23]. Despite being overcomplete and non-orthogonal, the basis formed by coherent states is useful whenever one needs to calculate *normally ordered* combinations of operators, i. e., combinations where all creation operators stand on the left and all annihilation operators stand on the right. This is because a normally ordered operator is averaged over a coherent state in a simple way using Eqs (11.16) and (11.17):

$$\langle\alpha|(a^{\dagger})^m (a)^n|\alpha\rangle = (\alpha^*)^m \alpha^n. \tag{11.19}$$

In particular, the mean photon number of a coherent state is $\langle\alpha|\hat{N}|\alpha\rangle = |\alpha|^2$.

A good approximation to a coherent state is the state generated by a laser [21]. Although from the viewpoint of classical optics, laser radiation has constant intensity [15], a quantum coherent state has fluctuations of the photon number: *shot noise*, a consequence of the discrete structure of light. Indeed, the variance of the photon number in a coherent state, defined as $\mathrm{Var}(N_{\mathrm{coh}}) \equiv \langle \alpha | (\hat{N} - \langle \alpha | \hat{N} | \alpha \rangle)^2 | \alpha \rangle$, is

$$\mathrm{Var}(N_{\mathrm{coh}}) = |\alpha|^2 = \langle N \rangle_{\mathrm{coh}}. \tag{11.20}$$

The *uncertainty* of an observable is estimated as the square root of its variance. Accordingly, the uncertainty given by the shot noise is $\Delta N_{\mathrm{coh}} = \sqrt{\langle N \rangle_{\mathrm{coh}}}$.

The variances of the quadrature operators are also nonzero in a coherent state. From the definitions of the quadrature operators (11.9), it follows that

$$\mathrm{Var}(q_{\mathrm{coh}}) = \mathrm{Var}(p_{\mathrm{coh}}) = \frac{1}{4}. \tag{11.21}$$

Therefore, any coherent state has the same uncertainty of any quadrature: $\Delta q = \Delta p = \Delta q_\varphi = \frac{1}{2}$. The mean values of the quadratures in a coherent state are $\langle \alpha | \hat{q} | \alpha \rangle = \mathfrak{Re}\{\alpha\}$ and $\langle \alpha | \hat{p} | \alpha \rangle = \mathfrak{Im}\{\alpha\}$.

A real-world coherent state differs from this idealized picture. First, it has inevitable phase fluctuations due to the phase drift of the laser. This problem can be overcome in experiments by using the same laser as the source of a quantum state and as a reference. Because what matters is the relative phase between the state and the reference, the phase drift does not affect the measurement. The second problem is the excess noise in the number of photons, which makes the uncertainty in the photon number larger than the shot noise. The photon-number variance due to the excess noise scales quadratically with the mean photon number; due to this fact, the role of excess noise reduces as the mean photon number decreases. Therefore, no laser is *shot-noise limited*, i. e., has the photon-number uncertainty given by the shot noise; however, any laser can be made shot-noise limited by sufficiently reducing its intensity.

The *vacuum state* $|0\rangle$ belongs to both the Fock states and the coherent states. As a Fock state, it has the eigenvalue $N = 0$, and this photon number does not fluctuate. Meanwhile, as a coherent state, it has nonzero uncertainties of the quadratures. This shows that a quantum vacuum is not just 'nothing': despite having zero mean number of photons and zero mean field, it has nonzero field fluctuations (*zero-point fluctuations of the electromagnetic vacuum*).

Squeezed states are formally defined as the eigenstates of the operator $\mu a + \nu a^\dagger$ [30]. More commonly, they are known as the states in which the uncertainty of one quadrature is smaller than of the other one, for instance, $\Delta q < \Delta p$. At the same time, these are minimal-uncertainty states, i. e., the product of their quadrature uncertainties is $\Delta q \Delta p = 1/4$, as in the case of coherent states. It follows that the uncertainty of some quadrature $\hat{q}_\varphi$ is smaller than the shot-noise uncertainty, for instance,

$\Delta q < 1/2$. One says that the quantum fluctuations are *squeezed* for this quadrature and *anti-squeezed* for the conjugated one, $\Delta p > 1/2$. The reduced noise in the squeezed quadrature makes squeezed states useful for metrology. In particular, a *squeezed-vacuum state* has $\langle \hat{q} \rangle = \langle \hat{p} \rangle = 0$, but unequal uncertainties of different quadratures. Due to the increased noise in the anti-squeezed quadrature, a squeezed vacuum state has a mean number of photons that is not only nonzero, but sometimes very large.

11.1.3 State characterization

To characterize and distinguish various quantum states, several instruments are used. Here we will consider two of them: Glauber's correlation functions and the Wigner function. Further, we will apply these instruments to characterize nonclassical states of polarized light.

Glauber's correlation functions of order n describe n-photon absorption. In particular, they determine the outcome of the Hanbury Brown–Twiss experiment [15, 23] where an incident beam is split, in general, into n beams, with a detector in each beam, and the simultaneous photocounts of these detectors are registered. The photocount coincidence rate or, in the case of bright light, the correlation of the photocurrents, is given by the normally ordered nth-order correlation function. Formally, it is defined as [15]

$$G^{(n)}(\vec{r}_1, t_1; \ldots; \vec{r}_n, t_n)$$
$$\equiv \langle \hat{E}^{(-)}(\vec{r}_1, t_1) \cdots \hat{E}^{(-)}(\vec{r}_n, t_n) \hat{E}^{(+)}(\vec{r}_n, t_n) \cdots \hat{E}^{(+)}(\vec{r}_1, t_1) \rangle, \qquad (11.22)$$

where $\vec{r}_1, \ldots \vec{r}_n$ and $t_1 \ldots t_n$ are the positions and times at which the detectors measure and the averaging is over the quantum state. Note that the definition assumes normal ordering, i. e., all negative-frequency operators standing on the left and positive-frequency operators, on the right.

Because correlation functions (11.22) depend on the mean number of photons, it is convenient to normalize them as

$$g^{(n)}(\vec{r}_1, t_1; \ldots; \vec{r}_n, t_n)$$
$$\equiv \frac{\langle \hat{E}^{(-)}(\vec{r}_1, t_1) \cdots \hat{E}^{(-)}(\vec{r}_n, t_n) \hat{E}^{(+)}(\vec{r}_n, t_n) \cdots \hat{E}^{(+)}(\vec{r}_1, t_1) \rangle}{\langle \hat{E}^{(-)}(\vec{r}_1, t_1) \hat{E}^{(+)}(\vec{r}_1, t_1) \rangle \cdots \langle \hat{E}^{(-)}(\vec{r}_n, t_n) \hat{E}^{(+)}(\vec{r}_n, t_n) \rangle}. \qquad (11.23)$$

Normalized correlation functions are convenient tools to characterize the photon statistics. For instance, for a coherent state $g^{(n)} = 1$, for a thermal state $g^{(n)} = n!$.

Most commonly used is the second-order correlation function,

$$g^{(2)}(\vec{r}_1, t_1; \vec{r}_2, t_2)$$
$$\equiv \frac{\langle \hat{E}^{(-)}(\vec{r}_1, t_1) \hat{E}^{(-)}(\vec{r}_2, t_2) \hat{E}^{(+)}(\vec{r}_2, t_2) \hat{E}^{(+)}(\vec{r}_1, t_1) \rangle}{\langle \hat{E}^{(-)}(\vec{r}_1, t_1) \hat{E}^{(+)}(\vec{r}_1, t_1) \rangle \langle \hat{E}^{(-)}(\vec{r}_2, t_2) \hat{E}^{(+)}(\vec{r}_2, t_2) \rangle}. \qquad (11.24)$$

In the case of a stationary spatially homogeneous radiation, $g^{(2)}$ depends only on the time delay $t_1 - t_2 \equiv \tau$ and space displacement $\vec{r}_1 - \vec{r}_2 \equiv \vec{\rho}$. Its value at $\tau = 0, \vec{\rho} = 0$ is known as the *bunching parameter*, and it can be written in terms of photon-creation and -annihilation operators as

$$g^{(2)}(0) = \frac{\langle (a^\dagger)^2 a^2 \rangle}{\langle a^\dagger a \rangle^2} \equiv \frac{\langle : \hat{N}^2 : \rangle}{\langle \hat{N} \rangle^2}. \tag{11.25}$$

Here the notation ':' means normal ordering.

Equations (11.22)–(11.25) describe the correlation functions for a single radiation mode. The study of polarized nonclassical light also requires two-mode correlation functions. For instance, the second-order cross-correlation function for horizontal and vertical polarization modes at zero delay and displacement is

$$g^{(2)}_{H,V}(0) = \frac{\langle a_H^\dagger a_V^\dagger a_H a_V \rangle}{\langle a_H^\dagger a_H \rangle \langle a_V^\dagger a_V \rangle} \equiv \frac{\langle \hat{N}_H \hat{N}_V \rangle}{\langle \hat{N}_H \rangle \langle \hat{N}_V \rangle}. \tag{11.26}$$

The normal ordering is omitted here because the operators of orthogonal polarization modes commute.

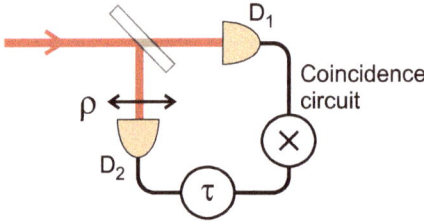

Figure 11.2: Measurement of the Glauber correlation function and the bunching parameter in the Hanbury Brown–Twiss experiment.

The second-order correlation function (11.24) can be measured in a Hanbury Brown–Twiss setup (Fig. 11.2). The state of light under study is sent to a beamsplitter, and the radiation in its output ports is registered by two detectors D_1, D_2. In the case of 'faint' light, single-photon detectors should be used, and their output pulses are sent to a coincidence circuit or to a time-tagger. Counts of the two detectors are considered coincident if they appear within the coincidence resolution time T_c. The correlation function is calculated as

$$g^{(2)}(\tau, \vec{\rho}) = \frac{R_c}{R_1 R_2 T_c}, \tag{11.27}$$

where R_c is the coincidence rate and R_1, R_2 are the rates of counts in the two detectors [23]. The time delay τ can be introduced electronically and the spatial displacement $\vec{\rho}$, by shifting one of the detectors. In the case of pulsed light, the equation for calculating $g^{(2)}$ is modified and, in general, contains both the coincidence resolution

T_c and the pulse duration. However, if the pulse is much shorter than T_c, the bunching parameter can be calculated using a simplified formula [11]:

$$g^{(2)}(0) = \frac{N_c}{N_1 N_2},$$

(11.28)

where N_c is the mean number of coincidences per pulse and N_1, N_2 are the mean numbers of photocounts per pulse in the two detectors.

The denominators in Eqs. (11.27) and (11.28) are equal, respectively, to the rate and mean number per pulse of coincidences in the case where detectors 1, 2 register light from independent sources (*accidental coincidences*). For coherent light, the arrival of each photon is independent of the others, all coincidences are accidental, and $g^{(2)}(0) = 1$. For single-mode thermal light, the number of coincidences is twice as large, yielding $g^{(2)}(0) = 2$. This result of the Hanbury Brown–Twiss experiment [15, 23] was interpreted as 'bunching' of photons [21] in thermal light—this is where the term 'bunching parameter' comes from. But this result also follows from the classical description of intensity fluctuations in thermal light [15]. What indeed requires a quantum description is *anti-bunching*, i. e., the case of $g^{(2)}(0) < 1$. Anti-bunching of photons can be observed for Fock states; in particular, for a single-photon state $g^{(2)}(0) = 0$.

For measuring the cross-correlation function (11.26), the beamsplitter in Fig. 11.2 should be a polarization one. The Hanbury Brown–Twiss setup can be also extended to the general case of measuring the nth-order correlation function. For this, one should use n detectors after a sufficient number of beamsplitters, and register n-fold, instead of two-fold, coincidences.

Quasi-probabilities. In order to describe a state in terms of quadratures q and p, i. e., in the *phase space*, it would be very convenient to have some joint probability distribution $P(q, p)$. But because the quadrature operators do not commute, their joint probability distribution is unphysical. Several quasi-probabilities can be introduced, but in each case there is a price to pay: the quasi-probabilities violate certain rules that normal probabilities should obey.

In classical probability theory, a probability distribution has a Fourier transform, called the characteristic function. Similarly, quantum quasi-probabilities can be defined as Fourier transforms of certain characteristic functions. The normally ordered characteristic function is defined as [23]

$$C_n(w) = \langle e^{wa^\dagger} e^{-w^* a} \rangle,$$

(11.29)

where w is a complex number and the averaging is over the quantum state to be characterized. The Fourier transform of $C_n(w)$ is the *Glauber–Sudarshan quasi-probability, or P-distribution*:

$$P(z) = \frac{1}{\pi^2} \int d^2 w C_n(w) e^{-wz^* + w^* z},$$

(11.30)

with $z = q + ip$ being a complex number. The P-distribution has the meaning of the density matrix in the coherent-state representation. It can be singular or negative for some states—this is why it is not a true probability distribution. In fact, its negativity is a criterion of nonclassicality: by definition, *a nonclassical state is one that has a negative Glauber–Sudarshan quasi-probability $P(q,p)$*. The negativity of the P-distribution means that a state cannot be described in terms of classical statistical optics [30]. However, because $P(q,p)$ can be singular, it cannot be measured directly. For the measurement, most convenient is the Wigner function.

The Wigner function is defined as the two-dimensional Fourier transform of the symmetrized characteristic function [23]:

$$W(z) = \frac{1}{\pi^2} \int d^2 w C_s(w) e^{-wz^* + w^* z}. \tag{11.31}$$

The symmetrized characteristic function $C_s(w)$ is

$$C_s(w) = \langle e^{wa^\dagger - w^* a} \rangle. \tag{11.32}$$

The Wigner function cannot be singular, but it can be negative. From the negativity of the Wigner function, the negativity of the P-distribution follows; in other words, the negativity of the Wigner function is a sufficient condition for nonclassicality.

The most important property of the Wigner function is that it can be used for calculating the mean values and moments of a quadrature in the same manner as a 'normal' probability distribution:

$$\langle q^n \rangle = \int q^n W(q,p) dq dp. \tag{11.33}$$

This feature means that the marginal distribution $W_q(q) \equiv \int W(q,p) dp$ is a true probability distribution. The same property is valid for the marginal distribution of any generalized quadrature $q_\varphi \equiv q \cos\varphi + p \sin\varphi$. It also provides a way to measure the Wigner function in experiment [20].

Measurement of the Wigner function is performed through *balanced homodyne detection* (Fig. 11.3). A state to be characterized is sent to a beamsplitter, as in the Hanbury Brown–Twiss experiment, but now another state is sent into the second input port, namely, a strong coherent state known as the *local oscillator* [2]. Importantly, the beamsplitter should be perfectly balanced (50 %), and the local oscillator should be much brighter than the state under study. The detectors should then not be counting photons but registering strong photon fluxes. At the output, their photocurrents $i_{1,2}$ should be subtracted.

The difference photocurrent, $i_- \equiv i_1 - i_2$, is then proportional to the quadrature of the input state,

$$i_- = 2\eta\alpha_0 q_\varphi \tag{11.34}$$

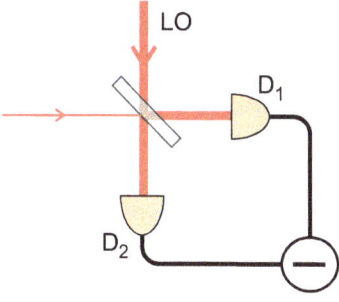

Figure 11.3: Balanced homodyne detection.

where η is the quantum efficiency of the detectors (assumed to be the same), α_0 the amplitude of the local oscillator, and φ its phase. Equation (11.34) can be used to measure the mean value and all statistical moments of the quadrature q_φ, by analyzing the probability distribution of the difference photocurrent i_-. Such distributions can be acquired for a set of quadratures $\{q_\varphi\}$ with different φ, and then, using the property (11.33), one can reconstruct the Wigner function through the inverse Radon transformation or some other method. This procedure is called the *Wigner-function tomography*.

It is worth noting that the scheme in Fig. 11.3 resembles the Stokes measurement setup (Fig. 5.7), in that the value of interest is obtained by subtracting the photocurrents of two detectors. This analogy will be further developed in Section 11.3.

Observable signs of nonclassicality. Although the rigorous definition of nonclassical light is in terms of the P-distribution and cannot be applied in experiment, there are many observable features that follow from the negativity of the P-function and therefore are sufficient conditions for nonclassicality. Some of them were briefly mentioned above, but here we present a more complete (but not exhaustive) list of *nonclassicality signs* used in experiment.

1. Anti-bunching, $g^{(2)}(0) < 1$. This feature, already mentioned above, is equivalent to another property, namely, *sub-Poissonian statistics*. Because the bunching parameter is related to the variance and mean of the photon number as

$$g^{(2)}(0) - 1 = \frac{\text{Var}(N) - \langle N \rangle}{\langle N \rangle^2}, \tag{11.35}$$

anti-bunching means that the variance is less than the mean, $\text{Var}(N) < \langle N \rangle$. However, experimental conditions for observing anti-bunching and sub-Poissonian statistics are different: according to Eq. (11.35), anti-bunching is easier to detect for faint light, $\langle N \rangle < 1$, while sub-Poissonian statistics is better noticeable for bright light, $\langle N \rangle \gg 1$.

2. Anti-bunching can be generalized to the condition involving higher-order normalized correlation functions $g^{(n)} \equiv g^{(n)}(0)$ [13]. The resulting sufficient conditions for nonclassicality are

$$\frac{g^{(n-1)}g^{(n+1)}}{[g^{(n)}]^2} < 1. \tag{11.36}$$

3. Several nonclassicality conditions are formulated in terms of the probability distribution $p(m)$ for the photocount number m. One of them is [13]

$$\frac{m+1}{m}\frac{p(m-1)p(m+1)}{[p(m)]^2} < 1. \tag{11.37}$$

Efficient conditions have been derived for single-photon ('on/off') detectors, in terms of probabilities to detect one or no photons $p(0)$ [10, 29].

4. Quadrature squeezing, already mentioned above, i. e., the uncertainty of one quadrature q_φ being less than the vacuum noise $1/2$, is also a sufficient condition for nonclassicality.

5. The 'strongest' sign of nonclassicality is the negativity of the Wigner function. Experimentally, this is most difficult condition to witness. According to Hudson's theorem [20], the Wigner function of a pure state is negative if and only if it is non-Gaussian.

So far, we listed sufficient conditions for the nonclassicality of a state in a single mode. Further, we continue with states defined in two modes, including polarization modes.

6. *Twin-beam squeezing* is an effect where the variance of photon-number difference in two modes 1, 2 is smaller than the sum of mean photon numbers in these modes,

$$\mathrm{Var}(N_1 - N_2) < \langle N_1 + N_2 \rangle. \tag{11.38}$$

Twin-beam squeezing violates the classical Cauchy–Schwarz inequality [31] and is therefore a sufficient condition of nonclassicality. In the framework of this book, most important is *polarization squeezing* [7], where 1, 2 are polarization modes:

$$\mathrm{Var}(N_H - N_V) < \langle N_H + N_V \rangle. \tag{11.39}$$

7. *Entanglement.* A state $|\Psi\rangle_{1,2}$ of two quantum systems, or two radiation modes 1, 2, is entangled if it cannot be represented as a product of states in these modes:

$$|\Psi\rangle_{12} \neq |\Psi\rangle_1 |\Psi\rangle_2. \tag{11.40}$$

Otherwise, the state $|\Psi\rangle_{1,2}$ is called *separable*. A mixed state ρ_{12} is called separable if its density matrix can be represented as a convex sum of factorizable states ρ_1, ρ_2 in modes 1, 2; otherwise it is called *inseparable*.

To certify entanglement in experiment, there are various witnesses and measures, which, however, will not be used in this book.

11.2 Stokes observables

As photon-number operators are quantum counterparts of intensities in classical optics, the classical Stokes observables (see Chapter 3) are naturally replaced in quantum optics by the Stokes operators [7]:

$$\hat{S}_0 \equiv a_H^\dagger a_H + a_V^\dagger a_V,$$
$$\hat{S}_1 \equiv a_H^\dagger a_H - a_V^\dagger a_V,$$
$$\hat{S}_2 \equiv a_H^\dagger a_V + a_V^\dagger a_H,$$
$$\hat{S}_3 \equiv -i(a_H^\dagger a_V - a_V^\dagger a_H). \tag{11.41}$$

The Stokes operators are Hermitian, by definition, and they correspond to real observables. Furthermore, we will consider the measurement of these observables, but from now on we will use the term 'Stokes parameters' exclusively for their mean values. As in classical optics, it is worth introducing a Stokes operator of a general form,

$$\hat{S}(\vartheta, \varphi) \equiv \hat{S}_1 \cos\vartheta + \hat{S}_2 \sin\vartheta \cos\varphi + \hat{S}_3 \sin\vartheta \sin\varphi. \tag{11.42}$$

Similarly to definition (3.30) in classical polarization optics, in quantum optics the degree of polarization is defined as [7]

$$P \equiv \frac{\sqrt{\langle \hat{S}_1 \rangle^2 + \langle \hat{S}_2 \rangle^2 + \langle \hat{S}_3 \rangle^2}}{\langle \hat{S}_0 \rangle}. \tag{11.43}$$

Because this definition is insufficient to describe certain effects, other definitions have been proposed [17]. We will consider them in Chapter 12.

11.2.1 Commutation and uncertainty relations

Generally, the Stokes operators (11.41) do not commute. Their commutation relations are obtained from definitions (11.41), with the help of the rules

$$[\hat{A}, \hat{B} + \hat{C}] = [\hat{A}, \hat{B}] + [\hat{A}, \hat{C}],$$
$$[\hat{A}, \hat{B}\hat{C}] = [\hat{A}, \hat{B}]\hat{C} + \hat{B}[\hat{A}, \hat{C}]. \tag{11.44}$$

As a result, the commutation relations for the Stokes operators read

$$[\hat{S}_0, \hat{S}_1] = [\hat{S}_0, \hat{S}_2] = [\hat{S}_0, \hat{S}_3] = 0,$$
$$[\hat{S}_1, \hat{S}_2] = 2i\hat{S}_3,$$
$$[\hat{S}_2, \hat{S}_3] = 2i\hat{S}_1,$$
$$[\hat{S}_3, \hat{S}_1] = 2i\hat{S}_2. \tag{11.45}$$

These commutation relations resemble the ones of the Pauli operators. This again points at the analogy between polarized photons and a spin 1/2 particle, which was already mentioned in Chapter 6. In what follows, we will further develop this analogy.

From Eqs. (11.45), the uncertainty relations follow. Indeed, one can show [19] that the uncertainties of non-commuting operators $\hat{A}$ and $\hat{B}$, defined as $\Delta A \equiv \sqrt{\langle(\hat{A} - \langle\hat{A}\rangle)^2\rangle}$ and $\Delta B \equiv \sqrt{\langle(\hat{B} - \langle\hat{B}\rangle)^2\rangle}$, satisfy the condition

$$\Delta A \Delta B \geq \frac{1}{2}|\langle[\hat{A}, \hat{B}]\rangle|. \tag{11.46}$$

Accordingly, the Stokes operators satisfy the uncertainty relations

$$\Delta S_1 \Delta S_2 \geq |\langle\hat{S}_3\rangle|,$$
$$\Delta S_2 \Delta S_3 \geq |\langle\hat{S}_1\rangle|,$$
$$\Delta S_3 \Delta S_1 \geq |\langle\hat{S}_2\rangle|. \tag{11.47}$$

Physically, it means that different Stokes observables cannot be measured simultaneously unless one of the three has a zero mean value.

From definitions (11.41) of the Stokes operators, one can derive a useful identity [14],

$$\hat{S}_1^2 + \hat{S}_2^2 + \hat{S}_3^2 = \hat{S}_0(\hat{S}_0 + 2). \tag{11.48}$$

This identity leads to another inequality for the Stokes observables. Indeed, the definition (11.43) of the degree of polarization can be rewritten as

$$\langle\hat{S}_1\rangle^2 + \langle\hat{S}_2\rangle^2 + \langle\hat{S}_3\rangle^2 = P\langle\hat{S}_0\rangle^2.$$

Subtracting this equation from Eq. (11.48), averaging and using the conditions $0 \leq P \leq 1$, $\mathrm{Var}(\hat{S}_0) \geq 0$, we obtain the relation

$$\mathrm{Var}(\hat{S}_1) + \mathrm{Var}(\hat{S}_2) + \mathrm{Var}(\hat{S}_3) \geq 2\langle\hat{S}_0\rangle. \tag{11.49}$$

Inequality (11.49) is sometimes interpreted as an additional uncertainty relation.

Further, we will consider the quantum-mechanical approach to the measurement of the Stokes observables. But before, let us introduce a class of states for which this measurement will be described.

11.2.2 A single-photon state

In quantum optics, to measure the mean value of some observable A means to find the average of the corresponding operator $\hat{A}$ over a state. If the averaging is over a pure state $|\Psi\rangle$, it is defined as

$$\langle\hat{A}\rangle \equiv \langle\Psi|\hat{A}|\Psi\rangle. \tag{11.50}$$

The averaging over a mixed state with the density matrix $\hat{\rho}$ is written as

$$\langle\hat{A}\rangle \equiv \mathrm{Tr}(\hat{A}\hat{\rho}). \tag{11.51}$$

To describe the quantum measurement of the Stokes observables, we will consider a single-photon state, which is a generic superposition of $N = 1$ Fock states in the horizontal and vertical polarization modes:

$$|\Psi\rangle = \alpha|1\rangle_H|0\rangle_V + \beta|0\rangle_H|1\rangle_V \equiv \alpha|1\rangle_H + \beta|1\rangle_V. \tag{11.52}$$

Further, we will use the last notation for brevity. With the normalization condition $|\alpha|^2 + |\beta|^2 = 1$, the state (11.52) describes a single photon that is 'spread' over the two polarization modes H, V. The state can be also represented as a two-component vector,

$$|\Psi\rangle = \begin{pmatrix} \alpha \\ \beta \end{pmatrix}, \tag{11.53}$$

which resembles the Jones vector (3.4) we considered in Chapter 3. In particular, the common phase of the coefficients α and β plays no role now.

The state (11.53) represents a general state of a *qubit*, the state of a quantum system with two eigenstates. This can be a two-level atom, a spin 1/2 particle like an electron, or—as we see—a polarized photon. The latter, as a result, can represent any of these other quantum systems.

Some particular cases of a polarized single photon are as follows. The states

$$|1\rangle_H, \quad |1\rangle_V \tag{11.54}$$

are horizontally and vertically polarized photons; the states

$$|1\rangle_D = \frac{1}{\sqrt{2}}(|1\rangle_H + |1\rangle_V), \quad |1\rangle_A = \frac{1}{\sqrt{2}}(|1\rangle_H - |1\rangle_V) \tag{11.55}$$

are diagonally and anti-diagonally polarized photons, and

$$|1\rangle_L = \frac{1}{\sqrt{2}}(|1\rangle_H + i|1\rangle_V), \quad |1\rangle_R = \frac{1}{\sqrt{2}}(|1\rangle_H - i|1\rangle_V) \tag{11.56}$$

are left- and right-circularly polarized photons.

11.2.3 Quantum measurement of the Stokes observables

The 'canonical' definition of measurement in quantum mechanics is through a projection on the eigenstates of the corresponding operator (*von Neumann, or projective, measurement)* [24]. For instance, above we saw that a Fock state is an eigenstate of the photon-number operator. A generic Hermitian operator $\hat{A}$ always has a complete orthonormal set of eigenstates: $\hat{A}|A_n\rangle = A_n|A_n\rangle$. Any state $|\Phi\rangle$ can be decomposed over this set,

$$|\Phi\rangle = \sum_n c_n|A_n\rangle, \tag{11.57}$$

where $c_n = \langle A_n|\Phi\rangle$. The mean value of $\hat{A}$ over $|\Phi\rangle$ is found according to Eq. (11.50),

$$\langle\hat{A}\rangle = \langle\Phi|\hat{A}|\Phi\rangle = \sum_n |c_n|^2 A_n, \tag{11.58}$$

where we used the orthogonality of the eigenstates $|A_n\rangle$, $\langle A_n|A_m\rangle = \delta_{mn}$. The nth term in Eq. (11.58) is the probability that the state is $|A_n\rangle$, $P_n \equiv |c_n|^2 = |\langle\Phi|A_n\rangle|^2$, times the value of the operator $\hat{A}$ in this state. This expression is perfectly clear from the viewpoint of the probability theory: with the probability P_n, observable A takes the value A_n.

The same result is achieved using the decomposition of the identity over the eigenstates of $\hat{A}$. Indeed, by plugging the decomposition

$$\hat{\mathbf{1}} = \sum_n |A_n\rangle\langle A_n| \tag{11.59}$$

between $\langle\Phi|\hat{A}$ and $|\Phi\rangle$ in Eq. (11.58), we obtain

$$\langle\hat{A}\rangle = \langle\Phi|\sum_n \hat{A}|A_n\rangle\langle A_n|\Phi\rangle = \sum_n A_n P_n. \tag{11.60}$$

In other words, the probability to measure a certain eigenvalue A_n of operator $\hat{A}$ is given by the squared projection of a state on the corresponding eigenstate $|A_n\rangle$.

To apply this procedure to the Stokes operators, we need to find their eigenstates and eigenvalues. We will do it for the general case of a polarized single photon (11.52).

First, we notice that the single-photon state (11.52) is an eigenstate for the operator $\hat{S}_0$, with the eigenvalue 1. To find the eigenstates and eigenvalues of the first Stokes operator $\hat{S}_1$, we have to solve the equation

$$\hat{S}_1|\Psi\rangle = s_1|\Psi\rangle. \tag{11.61}$$

Substituting (11.52) and taking into account that $\hat{N}_H|1\rangle_H = |1\rangle_H$, $\hat{N}_V|1\rangle_V = |1\rangle_V$, we get

$$\alpha(1 - s_1)|1\rangle_H = \beta(1 + s_1)|1\rangle_V.$$

Because the Fock states $|1\rangle_H$ and $|1\rangle_V$ belong to orthogonal modes, the factors in front of them should be zero. From this, we find two possibilities: either $s_1 = 1$ and $\beta = 0$, or $s_1 = -1$ and $\alpha = 0$.

We obtained the result that the eigenstates of $\hat{S}_1$ are horizontally polarized single photon and vertically polarized single photon, and the corresponding eigenvalues are +1 and −1.

For the second Stokes operator $\hat{S}_2$, the eigenvalue problem leads to the equation

$$\hat{S}_2|\Psi\rangle = s_2|\Psi\rangle. \tag{11.62}$$

To solve this equation, we notice that the second Stokes operator converts horizontally and vertically polarized photons into each other: $\hat{S}_2|1\rangle_H \equiv (a_H^\dagger a_V + a_V^\dagger a_H)|1\rangle_H = |1\rangle_V$, and similarly $\hat{S}_2|1\rangle_V = |1\rangle_H$. Then the equation becomes

$$(\alpha - s_2\beta)|1\rangle_V = (s_2\alpha - \beta)|1\rangle_H.$$

Again, requiring that the factors by the states $|1\rangle_H$ and $|1\rangle_V$ are both zero, we obtain two solutions: either $s_2 = 1$ and $\alpha = \beta = 1/\sqrt{2}$, or $s_2 = -1$ and $\alpha = -\beta = 1/\sqrt{2}$. Thus, the eigenstates of $\hat{S}_2$ are a diagonally polarized single photon, with the eigenvalue +1, and an anti-diagonally polarized single photon, with the eigenvalue −1.

Similarly, one can show that the eigenstates of $\hat{S}_3$ are a left-circularly polarized single photon, with the eigenvalue +1, and a right-circularly polarized single photon, with the eigenvalue −1.

For measuring a certain Stokes observable for a single photon, one should project it on the corresponding eigenstate. This is done with the same setup as in classical optics (Figs. 3.4, 3.5), with the only difference that the detectors should be able to register single photons (single-photon, or 'click' detectors). For instance, to measure S_1 we use the setup shown in Fig. 3.4. If the upper detector clicks, we say that the photon is horizontally polarized, and write down the result: $s_1 = 1$. If the lower detector clicks, our result is $s_1 = -1$. This procedure is actually very similar to using the formula $S_1 \equiv I_H - I_V$ (Section 3.3.4), with the intensities replaced by photon numbers: $S_1 \equiv N_H - N_V$. But for a single photon, either $N_H = 1$, $N_V = 0$ or the other way round, hence the Stokes observable takes values ±1.

After M tries *with identically prepared photons*, we calculate the mean value, i. e., the first Stokes parameter, by averaging the results of all tries:

$$\langle S_1 \rangle = \frac{1}{M} \sum_{i=1}^{M} s_{1i}, \tag{11.63}$$

where s_{1i} is the result of the ith try.

The variance of the first Stokes observable can be calculated in the same manner,

$$\text{Var}(\hat{S}_1) = \frac{1}{M} \sum_{i=1}^{M} s_{1i}^2 - \left[\frac{1}{M} \sum_{i=1}^{M} s_{1i} \right]^2. \tag{11.64}$$

This procedure of measuring $\hat{S}_1$ is very similar to the *Stern–Gerlach experiment* where the projection of the spin on a certain direction is measured for a spin-1/2 particle (Fig. 11.4(a)). In a magnetic field, the particle trajectory will bend, depending on the

a

b

c

d

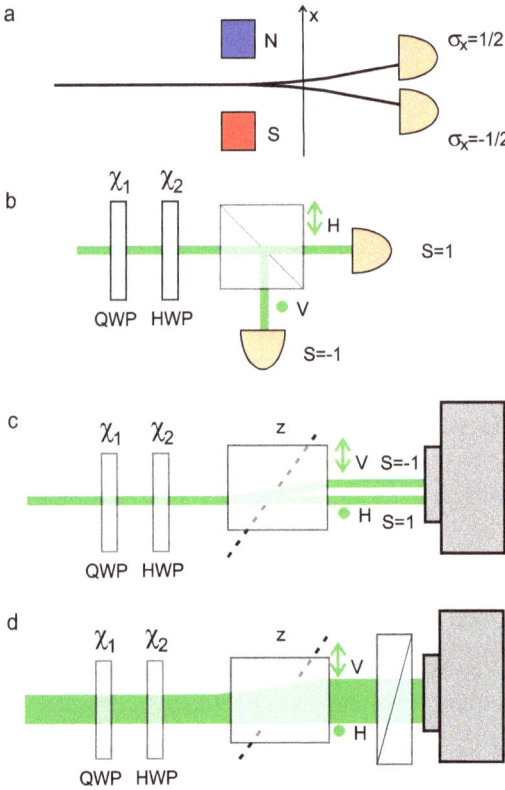

Figure 11.4: Projective measurement of the spin component σ_x for a spin-1/2 particle (a) and of a generic Stokes observable for a single photon (b), (c). Panel d shows the weak measurement of a generic Stokes observable.

projection of its spin $\vec{\sigma}$ on the direction of the magnetic field (x in the figure). Note that, for measuring the projection of $\vec{\sigma}$ on some other direction, for instance, y, the magnet in the figure should be rotated. It is impossible to measure σ_x and σ_y simultaneously because the corresponding Pauli operators do not commute.

The same is true for the Stokes observables. Indeed, although there is a setup for the measurement of any desired Stokes observable (Fig 5.7), it requires different settings for different Stokes observables. In this setup, shown for the case of quantum measurement in Fig. 11.4(b), a polarization prism is preceded by a QWP and a HWP. The orientation angles of the plates, χ_1 and χ_2, determine the parameters ϑ and φ of the generic Stokes observable (11.42). For instance, to measure S_1, both plates should be oriented horizontally: $\chi_1 = \chi_2 = 0°$. For the measurement of S_2, the angles should be $\chi_1 = 45°$ and $\chi_2 = 22.5°$, and for the measurement of $S_3, \chi_1 = 45°$ and $\chi_2 = 0°$. For general settings of the plates, the Stokes observable (11.42) measured in the setup is given by Eq. (5.25).

Clearly, the measurement of different Stokes observables S_1, S_2, S_3 requires different settings in the setup and therefore cannot be performed at once. In particular, if a photon is polarized diagonally or circularly, the measurement of S_1 will not give any information about its polarization state. In the setup of Fig. 11.4(b), the photon will be reflected or transmitted with 50 % probability. The fact that the photon is detected, say, in the 'transmitted path' only tells that it was not a vertically polarized photon. As we will see in Chapter 12, this feature underlies the principle of quantum key distribution with polarized photons.

The Stokes observables can be alternatively measured using the walk-off effect, as it was shown in Chapter 4. In this case, the photon should be detected after a long birefringent crystal, for instance, calcite (Fig. 4.8) with the optic axis in the vertical plane. If a photon is displaced in the course of propagation through the crystal, then it is vertically polarized, and the value $s_1 = -1$ is registered. If the photon is not displaced, the conclusion is that $s_1 = 1$. The measured value of s_1 should then correspond to the displacement Δx along the x axis: $s_1 = 1 - 2\Delta x/d$, where d is the shift due to the walk-off. It is assumed here that the beam size a is much smaller than d. A combination of quarter-wave and half-wave plates in front of the calcite turns the measurement of S_1 into the measurement of any other Stokes observable (Fig. 11.4(c)).

For $d \gg a$, the value of the vertical displacement, measured, for instance, with a camera, can be associated with the eigenvalue of $S_{\vartheta,\varphi}$. This condition makes the measurement projective (one also says 'strong'). In the next subsection we will see how the violation of this condition makes the measurement uncertain (one says 'weak') but, surprisingly, brings new interesting possibilities.

11.2.4 Weak measurement of the Stokes observables

Projective measurement enables unambiguous distinguishing between different eigenvalues, but at the same time it disturbs the quantum system very strongly. For instance, a photon after the experiment in Fig. 11.4(b), (c) will be polarized vertically or horizontally, no matter how it was polarized initially. However, it is possible to make a more 'gentle' but 'unprecise' measurement, where the measurement device disturbs the quantum system very little because their interaction is weak [4]. The amount of information extracted from such a measurement is also small, but the measurement can be repeated many times if necessary. In 1988, Aharonov et al. [1] showed that such a *weak measurement* performed over a spin-1/2 particle can result in absolute values of spin much exceeding 1/2. Later proposals [16] and experiments [25–27] were focused on measuring the Stokes observables, due to the similarity between the Stokes operators and the Pauli operators.

A Stokes observable for a photon will be measured weakly if the transverse shift in the 'walk-off' scheme (Fig. 11.4(c)) is very small compared to the initial uncertainty of the photon position, i. e., the beam size. This will be the case if the birefringent crystal is thin and the beam is broad (Fig. 11.4(d)). At the output, the transverse position of the photon is then disturbed very little, no matter what its initial polarization. It might seem that almost no information is gained; however, if the state is further projected on another state (with the help of a polarizer in Fig. 11.4(d)), the displacement can be made visible. The result of such a measurement is a so-called *weak value*, which can be larger than any of the Stokes eigenvalues [9].

The procedure of weakly measuring an observable A, corresponding to an operator $\hat{A}$, is mathematically described as follows. Let the initial state be $|\Psi^{in}\rangle$, a pure state for simplicity. First, observable A is measured weakly for this state. Then the resulting state is projected on the eigenstates of another operator $(\hat{B})$, which does not commute with $\hat{A}$. One of the eigenstates $|B_n\rangle$ is postselected.

Consider first the state $\hat{A}|\Psi^{in}\rangle$. Its decomposition over the eigenstates $|B_n\rangle$, according to Eq. (11.57), yields

$$\hat{A}|\Psi^{in}\rangle = \sum_n \langle B_n|\hat{A}|\Psi^{in}\rangle |B_n\rangle. \tag{11.65}$$

The mean value of $\hat{A}$ over $|\Psi^{in}\rangle$ is then found as

$$\langle \hat{A}\rangle \equiv \langle \Psi^{in}|\hat{A}|\Psi^{in}\rangle = \sum_n \langle B_n|\hat{A}|\Psi^{in}\rangle\langle \Psi^{in}|B_n\rangle, \tag{11.66}$$

which can be rewritten as

$$\langle \hat{A}\rangle = \sum_n |\langle B_n|\Psi^{in}\rangle|^2 \frac{\langle B_n|\hat{A}|\Psi^{in}\rangle}{\langle B_n|\Psi^{in}\rangle}. \tag{11.67}$$

This can be interpreted as the decomposition over 'weak values of A' [27],

$$\langle \hat{A}\rangle = \sum_n P_n A_n^w, \tag{11.68}$$

where

$$P_n = |\langle B_n|\Psi^{in}\rangle|^2 \tag{11.69}$$

is the probability and

$$A_n^w = \frac{\langle B_n|\hat{A}|\Psi^{in}\rangle}{\langle B_n|\Psi^{in}\rangle} \tag{11.70}$$

is the *weak value* of A.

If we choose a single outcome n of the experiment, for instance, $n = 0$, a certain weak value is postselected:

$$A_0^w \equiv \frac{\langle B_0|\hat{A}|\Psi^{in}\rangle}{\langle B_0|\Psi^{in}\rangle}. \tag{11.71}$$

The denominator in this expression can be very small if the output and input states are almost orthogonal. Then the weak value of A is very large. In particular, it can be larger than any of the eigenvalues A_n of the operator $\hat{A}$.

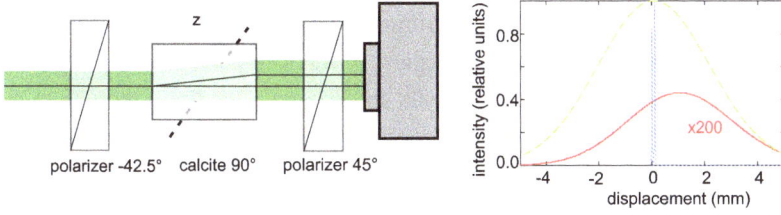

Figure 11.5: Left panel: weak measurement of the Stokes observable $\hat{S}_1$ for a state prepared by the first polarizer. Right panel: intensity distribution along the vertical axis without the second polarizer (green dashed line) and with it, scaled up by a factor of 200 (red solid line). Blue dotted line shows the intensity distribution for a narrow beam without the second polarizer (strong measurement).

Figure 11.5 shows how a weak value of a Stokes observable can be measured. The walk-off scheme, as in Fig. 11.4(d), performs the weak measurement of one Stokes observable and then a polarizer projects the state on another Stokes observable, not commuting with the first one. Suppose that the state at the input is a photon polarized almost anti-diagonally, $|\Psi^{in}\rangle = \alpha|1\rangle_H - \beta|1\rangle_V$, with $\alpha \approx \beta \approx 1/\sqrt{2}$. It is prepared by transmitting single photons through a polarizer oriented at an angle close to $-45°$. Then we weakly measure the observable $\hat{A} = \hat{S}_1$, with the help of a thin calcite crystal with the optic axis in the vertical plane. A camera placed close to the crystal (i. e., in the near field) measures the transverse displacement of the photon. This displacement is supposed to tell us the value of S_1, but for a broad input beam, it is hardly visible: the measurement is weak. However, before the camera we place a polarizer oriented at $45°$, and thus project the state onto an eigenstate of the $\hat{B} = \hat{S}_2$ operator, namely $|B_0\rangle = |1\rangle_D$. (Note that this state is nearly orthogonal to the input one.) Surprisingly, the displacement of the beam in the direction of the walk-off will be then very pronounced, and different for different input states. Indeed, the result of the measurement will be the weak value of operator $\hat{S}_1$ for the state $\alpha|1\rangle_H - \beta|1\rangle_V$, which is [see Eq. (11.71)]

$$\langle \hat{S}_1^w \rangle = \frac{{}_D\langle 1|\hat{S}_1(\alpha|1\rangle_H - \beta|1\rangle_V)}{{}_D\langle 1|(\alpha|1\rangle_H - \beta|1\rangle_V)} = \frac{\alpha + \beta}{\alpha - \beta}. \tag{11.72}$$

This value is very large if $\alpha \approx \beta$, and definitely larger than any eigenvalue of the operator $\hat{S}_1$. In other words, the displacement of the beam in the vertical direction will much exceed the displacement of a thin beam in a 'strong' measurement. At $\beta = 0$ or $\alpha = 0$, the weak value approaches the eigenvalues of $\hat{S}_1$ and the beam is displaced as in the 'strong' measurement.

The corresponding intensity distributions are shown in the right-hand panel of Fig. 11.5 for the case of a beam with full width at half maximum (FWHM) $a = 5$ mm and the displacement due to walk-off only $d = 0.1$ mm. The input state is linearly polarized at an angle $42.5°$, which corresponds to $\alpha = 0.74$, $\beta = 0.68$. In the absence of the second polarizer, the intensity distributions for the ordinary and extraordinary beams in calcite overlap (green dashed line) and cannot be distinguished. (For comparison, blue dotted line shows the intensity distributions for these beams if their widths are 0.01 mm, which is the case of strong measurement.) With the second polarizer inserted (red solid line), there is a single intensity peak, shifted about ten times more than in the case of a strong measurement. In other words, the small transverse shift due to the walk-off is amplified 10 times. In the limit of very small walk-off, the amplification factor tends to $\alpha/(\alpha - \beta) = 11.4$ [26].

This result has a simple classical explanation [26] in terms of the interference between the ordinary and extraordinary beams after the calcite. These beams overlap due to their large widths and have the same polarization states after the second polarizer. Their interference is destructive because the fields in the two beams have a π phase shift after the polarizer and very close absolute values. The resulting intensity peak is weak and shifted towards the stronger beam.

In a similar experiment, Salvail et al. [27] obtained weak values of the Stokes observables as large as 4, by projecting on states that were almost orthogonal to the input state. This shows how a weak measurement can retrieve very small displacements and therefore provides a precision higher than its strong counterpart.[1]

The weak value defined by Eq. (11.70) can be, in principle, also complex. It can be therefore used to directly measure the wavefunction of a quantum particle [22]. To this end, the near-field measurement discussed in this section should be complemented by a far-field measurement [27].

Finally, weak measurement can provide insights into the fundamental nonclassical features of light, such as the violation of Bell and Leggett–Garg inequalities and several quantum paradoxes [9].

11.3 Polarization quasi-probability

For the quantum description of polarized light, it is natural to introduce some quasi-probability distribution in the space of the Stokes observables S_1, S_2, S_3, similar to the quasi-probabilities in the phase space. Because the Stokes operators do not commute, this quasi-probability distribution is bound to have 'strange' features like negativity or singularity; however, it is still useful to describe the polarization part of the quantum state. Moreover, one could expect that its one- or two-dimensional marginals would have the properties of true probability distributions. These marginals can be helpful to develop some experimental state reconstruction procedure.

---

1 Strictly speaking, the precision is determined not only by the beam displacement, but also by its brightness (the number of photons). This taken into account, a weak measurement provides no advantage.

Indeed, in 2001 Karassiov and Masalov [5, 12] introduced the polarization quasi-probability as the Fourier transform of the symmetrized characteristic function

$$\chi(u_1, u_2, u_3) = \langle e^{u_1 \hat{S}_1 + u_2 \hat{S}_2 + u_3 \hat{S}_3} \rangle, \tag{11.73}$$

where $u_{1,2,3}$ are real Cartesian coordinates and the angular brackets denote the averaging over the quantum state, in the general case a mixed one.

The quasi-probability distribution, called the *polarization Wigner function*, is then defined as

$$W(S_1, S_2, S_3) = \iiint \chi(u_1, u_2, u_3) e^{-iu_1 S_1 - iu_2 S_2 - iu_3 S_3} \frac{du_1 du_2 du_3}{(2\pi)^3}. \tag{11.74}$$

For a certain generalized Stokes variable, $S_{\vartheta,\varphi} \equiv S_1 \cos\vartheta + S_2 \sin\vartheta \cos\varphi + S_3 \sin\vartheta \sin\varphi$, the polarization Wigner function leads to the marginal distribution [5]

$$P(S) = \iiint W(S_1, S_2, S_3) \delta(S_{\vartheta,\varphi} - S) dS_1 dS_2 dS_3. \tag{11.75}$$

Based on this relation, Bushev et al. implemented the tomographic reconstruction of the polarization quasi-probability distribution [5]. In this procedure, the state under study is sent into the classical setup for the measurement of the Stokes observables (Fig. 5.7). For a sufficient number of settings ϑ, φ, one should acquire the probability distributions (histograms) of the corresponding Stokes observables $S_{\vartheta,\varphi}$. These distributions are then processed with the help of the three-dimensional Radon transformation, which leads to the polarization quasi-probability distribution $W(S_1, S_2, S_3)$.

Figure 11.6 illustrates this procedure. Here the Poincaré sphere is shown in the space S_1, S_2, S_3, and not in the space of normalized Stokes observables σ_1, σ_2, σ_3 as in Fig. 5.3 and generally in Chapter 5. Because the photon number for the state, in the general case, is not fixed, only the Stokes space S_1, S_2, S_3 has a meaning. The sphere is

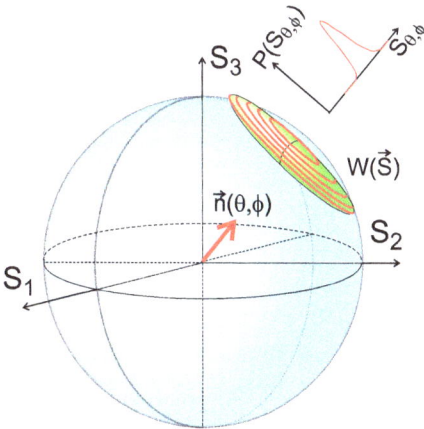

Figure 11.6: Polarization tomography: reconstruction of the 3D polarization quasi-probability distribution (green ellipsoid). Red lines show its cross-sections by planes orthogonal to the unit vector $\vec{n}(\vartheta, \varphi)$ defining the Stokes observable $S_{\vartheta,\varphi}$ to measure. The square areas of the cross-sections determine the probability distribution of this Stokes observable.

just 'a guide for the eye' and can indicate, for instance, the mean number of photons. The green ellipsoid shows schematically the surface where the quasi-probability distribution $W(S_1, S_2, S_3)$ has a certain value, for instance, its half-maximum value. For given angles ϑ, φ, determined by the orientations χ_1 and χ_2 of the HWP and QWP in Fig. 5.7 [see Eq. (5.25)], a single direction in the three-dimensional (3D) Stokes space, along the unit vector $\vec{n}(\vartheta, \varphi) = \{\cos\vartheta; \sin\vartheta\cos\varphi; \sin\vartheta\sin\varphi\}$, is probed. Each point of the histogram (11.75) is given by the surface area of the cross-section of the distribution $W(S_1, S_2, S_3)$ orthogonal to $\vec{n}(\vartheta, \varphi)$ at the corresponding value of $S_{\vartheta,\varphi}$ (shown by red lines).

This setup for polarization quantum tomography [5] strongly resembles the standard homodyne tomography setup (Fig. 11.3). In both cases, the settings of the setup (the phase of the local oscillator for homodyne tomography and the orientations of the HWP and QWP for polarization tomography) determine the direction in space: phase space for the Wigner-function tomography and the Stokes space for the polarization tomography. In both cases, the difference photocurrent of the two detectors is analyzed; its histogram determines the marginal probability distribution of the corresponding observable. But there is an important difference between the two schemes.

Unlike the Wigner function $W(q, p)$, the polarization quasiprobability distribution (11.74) is defined in terms of the Stokes observables, and the latter, unlike the quadratures, are integer-valued. This leads to the singularities and negative values even in the marginal distributions of the Stokes observables [6]. Even in the simplest case of a horizontally polarized weak coherent state, the calculated marginal probability $P(S_2, S_3)$ has singularities at integer values of $S_\perp \equiv \sqrt{S_2^2 + S_3^2}$ and is negative in the neighborhood of these values. The reconstructed quasi-probability distribution $W(S_1, S_2, S_3)$ will be also singular and negative at some values of S_1, S_2, S_3. This behavior can be observed through polarization tomography using the quantum Stokes measurement setup (Fig. 11.4(b)), where the detectors can count single photons. The resulting histogram of any Stokes observable will be discrete. After the reconstruction, $W(S_1, S_2, S_3)$ will show negative regions.

To demonstrate this, Spasibko et al. [28] reconstructed the probability distribution $P(S_2, S_3)$ for a coherent state with the mean number of photons 0.19. Sections of the polarization quasi-probability distribution by various planes and by the S_2 axis indeed showed negativities near the eigenvalues $S_2 = \pm 1$. The theory also predicts singularities at these points but they cannot be reconstructed from the experimental data.

Polarization tomography can be also performed with more advanced photon-number resolving detectors, which can register not only single photons but also multiphoton states and can distinguish between different photon numbers. An example is transition-edge sensors (TES). The use of such detectors would enable polarization quantum tomography in a larger space of photon numbers, but still, due to the discrete-valued experimental histograms, will result in a quasi-probability distribution with negativities.

Experiments with bright light, as the one of Ref. [5], require, instead of single-photon or photon-number resolving detectors, photocurrent detectors like p-i-n diodes. These detectors do not distinguish between different photon numbers and therefore smooth the histograms of the Stokes observables. The reconstructed polarization quasi-probability distribution will be free from singularities and negativities [6], as it was indeed the case in Ref. [5] and other experiments, described in more detail in Chapter 12. For example, the polarization quasi-probability distribution of a bright coherent state will be concentrated in the area around the point whose coordinates are the Stokes parameters for this state. The surface corresponding to its half-maximum value will be a sphere with the radius given by the uncertainty in the total photon number, i.e., $\Delta N = \sqrt{\langle N \rangle}$ [12]. For a generic bright state, the 'smoothed' polarization quasi-probability distribution is localized around the point $\langle S_1 \rangle; \langle S_2 \rangle; \langle S_3 \rangle$, and its typical sizes along the three axes in the Stokes space will be ΔS_1, ΔS_2, and ΔS_3.

We see that the negativity of the polarization Wigner function (11.74) cannot certify the nonclassicality of light: it can be both negative and singular but this is the case even for states whose P-distribution is positive, because the Stokes observables are integer-valued. Nevertheless, polarization quasi-probability distribution is a very useful tool because it can witness polarization squeezing, a nonclassical feature considered further in Chapter 12.

The analogy between the polarization quasi-probability and the Wigner function, between the polarization tomography and the homodyne tomography, can be developed further if we assume that the polarization quasi-probability distribution is localized far from the center of the Stokes space. This happens in experiments where a strong coherent state highlights one polarization mode, the other mode being occupied by some weaker quantum state. The polarization quasi-probability distribution will then be strongly shifted according to the polarization of the coherent component [3].

Suppose, for instance, that the coherent component is right-hand circularly polarized. It is then useful to consider just the part of the Poincaré sphere, whose radius is not unity but $\langle S_0 \rangle$, in the vicinity of the North Pole (Fig 11.7). If the circularly polarized coherent state is very bright, the Stokes parameter $\langle S_3 \rangle \gg 1$, and in the region where $W(S_1, S_2, S_3)$ is considerably nonzero the curvature of the Poincaré sphere is not very pronounced. If fluctuations of $\hat{S}_3$ are not targeted by the experiment (and they are mainly caused by the shot noise of the coherent state), only the plane (S_1, S_2) in the vicinity of the North Pole can be considered, sometimes called 'the dark plane' [8]. Under these conditions, polarization tomography reduces to the homodyne tomography, with the role of the local oscillator played by the strong circularly polarized coherent state [3]. The advantage of such a scheme is obvious: the local oscillator propagates in the same beam as the state under study, just in another polarization mode, and

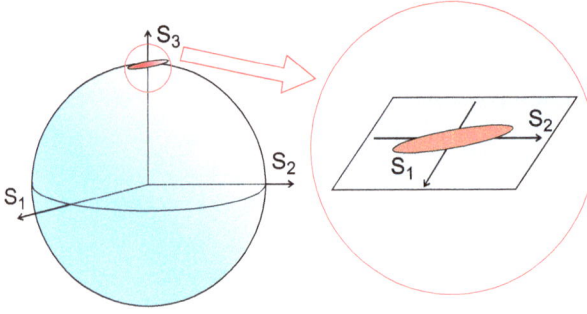

Figure 11.7: A state of polarized light with a strong circularly polarized coherent component. The Poincaré sphere has the radius $\langle S_0 \rangle$. The zoom shows the 'dark plane'.

therefore does not have to be phase locked. This also shows robustness of polarization tomography to phase drifts.

This situation is formally described as follows [18]. The photon-annihilation operators in modes H, V can be approximated by

$$a_{H,V} = \alpha_{H,V} + \delta \hat{q}_{H,V} + i\delta \hat{p}_{H,V}, \tag{11.76}$$

where $\alpha_H = \alpha_0$ and $\alpha_V = i\alpha_0$ are the amplitudes of the coherent states in the horizontal and vertical polarization modes, assumed to be classical, and $\delta \hat{q}_{H,V}$, $\delta \hat{p}_{H,V}$ are quadrature operators in the polarization modes H, V, which are assumed to be weakly populated. The Stokes operators $\hat{S}_{1,2}$ can then be expressed using definition (11.41) as

$$\hat{S}_1 = 2\alpha_0 (\delta \hat{q}_H - \delta \hat{p}_V),$$
$$\hat{S}_2 = 2\alpha_0 (\delta \hat{q}_V + \delta \hat{p}_H). \tag{11.77}$$

We see the Stokes operators $\hat{S}_1$, $\hat{S}_2$ scale as linear combinations of the quadrature operators in the H, V polarization modes, and the proportionality constant is given by the amplitude of the bright coherent state. The variances of the Stokes operators are also linearly related to the quadrature variances:

$$\mathrm{Var}(S_1) = 4|\alpha_0|^2 [\mathrm{Var}(q_H) + \mathrm{Var}(p_V)],$$
$$\mathrm{Var}(S_2) = 4|\alpha_0|^2 [\mathrm{Var}(q_V) + \mathrm{Var}(p_H)]. \tag{11.78}$$

Therefore, quadrature squeezing of the states populating the horizontally and vertically polarized modes will lead to the reduction in the fluctuations of the Stokes observables. This effect of *polarization squeezing* will be described in detail in the next chapter.

Bibliography

[1]   Y. Aharonov, D. Z. Albert, and L. Vaidman. How the result of a measurement of a component of the spin of a spin-1/2 particle can turn out to be 100. *Phys. Rev. Lett.*, 60:1351–1354, Apr 1988.

[2]   H.-A. Bachor and T. C. Ralph. *A guide to experiments in quantum optics*. Wiley-VCH, 2004.

[3]   W. P. Bowen, R. Schnabel, H. A. Bachor, and P. K. Lam. Polarization squeezing of continuous variable Stokes parameters. *Phys. Rev. Lett.*, 88:093601, Feb 2002.

[4]   V. B. Braginsky and F. Y. Khalili. *Quantum measurement*. Cambridge University Press, 1992.

[5]   P. A. Bushev, V. P. Karassiov, A. V. Masalov, and A. A. Putilin. Biphoton light with hidden polarization and its polarization tomography. *Opt. Spectrosc.*, 91:558–564, 2001.

[6]   M. V. Chekhova and F. Y. Khalili. Nonclassical features of the polarization quasiprobability distribution. *Phys. Rev. A*, 88:023822, Aug 2013.

[7]   A. S. Chirkin, A. A. Orlov, and D. Y. Parashchuk. Quantum theory of two-mode interactions in optically anisotropic media with cubic nonlinearities: Generation of quadrature- and polarization-squeezed light. *Rus. Journ. Quantum Electronics*, 23:870–874, 1993.

[8]   J. F. Corney, J. Heersink, R. Dong, V. Josse, P. D. Drummond, G. Leuchs, and U. L. Andersen. Simulations and experiments on polarization squeezing in optical fiber. *Phys. Rev. A*, 78:023831, Aug 2008.

[9]   J. Dressel, M. Malik, F. M. Miatto, A. Jordan, and R. W. Boyd. Colloquium: Understanding quantum weak values: basics and applications. *Rev. Mod. Phys.*, 86:307–316, 2014.

[10] R. Filip and L. Lachman. Hierarchy of feasible nonclassicality criteria for sources of photons. *Phys. Rev. A*, 88:043827, Oct 2013.

[11] O. A. Ivanova, T. S. Iskhakov, A. N. Penin, and M. V. Chekhova. Multiphoton correlations in parametric down-conversion and their measurement in the pulsed regime. *Quantum Electron.*, 36(10):951–956, oct 2006.

[12] V. P. Karassiov and A. V. Masalov. Quantum interference of light polarization states via polarization quasiprobability functions. *J. Opt. B, Quantum Semiclass. Opt.*, 4(4):S366–S371, aug 2002.

[13] D. N. Klyshko. The nonclassical light. *Phys. Usp.*, 39:573–596, 1996.

[14] D. N. Klyshko. Polarization of light: fourth-order effects and polarization-squeezed states. *J. Exp. Theor. Phys.*, 84:1065–1079, 1997.

[15] D. Klyshko. *Physical foundations of quantum electronics*. World Scientific, 2011.

[16] J. Knight and L. Vaidman. Weak measurement of photon polarization. *Phys. Lett. A*, 143:357–361, 1990.

[17] N. Korolkova. *Quantum polarization for continuous-variable information processing*, chapter 30, pages 405–417. John Wiley & Sons, Ltd, 2005.

[18] N. Korolkova, G. Leuchs, R. Loudon, T. C. Ralph, and C. Silberhorn. Polarization squeezing and continuous-variable polarization entanglement. *Phys. Rev. A*, 65:052306, Apr 2002.

[19] L. D. Landau and E. M. Lifshitz. *Quantum mechanics*. Elsevier, 3rd edition, 1977.

[20] U. Leonhardt. *Measuring the quantum state of light*. Cambridge University Press, 1997.

[21] R. Loudon. *The quantum theory of light*. Oxford University Press, 1973.

[22] J. S. Lundeen, B. Sutherland, A. Patel, C. Stewart, and C. Bamber. Direct measurement of the quantum wavefunction. *Nature*, 474:188–191, 2011.

[23] L. Mandel and E. Wolf. *Optical coherence and quantum optics*. Cambridge University Press, 1995.

[24] M. A. Nielsen and I. L. Chuang. *Quantum computation and quantum information*. Cambridge University Press, 2010.

[25] G. J. Pryde, J. L. O'Brien, A. G. White, T. C. Ralph, and H. M. Wiseman. Measurement of quantum weak values of photon polarization. *Phys. Rev. Lett.*, 94:220405, 2005.

[26] N. W. M. Ritchie, J. G. Story, and R. G. Hulet. Realization of a measurement of a 'weak value'. *Phys. Rev. Lett.*, 66:1107–1110, 1991.

[27] J. Z. Salvail, M. Agnew, A. S. Johnson, E. Bolduc, J. Leach, and R. W. Boyd. Full characterization of polarization states of light via direct measurement. *Nat. Photonics*, 7:316–321, 2013.

[28] K. Y. Spasibko, M. V. Chekhova, and F. Y. Khalili. Experimental demonstration of negative-valued polarization quasiprobability distribution. *Phys. Rev. A*, 96:023822, Aug 2017.

[29] J. Sperling, W. Vogel, and G. S. Agarwal. Correlation measurements with on–off detectors. *Phys. Rev. A*, 88:043821, Oct 2013.

[30] W. Vogel and D.-G. Welsch. *Quantum optics*. Wiley, 2006.

[31] D. F. Walls and G. J. Milburn. *Quantum optics*. Springer, 2008.

12 Nonclassical states of polarized light

Although in Chapter 11 we already considered single photons in an arbitrary polarization state, this chapter will give a broader overview of nonclassical states of polarized light and their features. These states are mainly prepared through nonlinear optical effects such as spontaneous parametric down-conversion (a second-order effect) and four-wave mixing or its analogues (third-order effects).

12.1 Spontaneous parametric down-conversion

Spontaneous parametric down-conversion (SPDC) can be viewed as reversed second-harmonic generation: while the latter is merging two photons at frequency ω into one photon at frequency 2ω, the former is splitting a single photon into a pair of photons. But unlike second-harmonic generation, spontaneous parametric down-conversion cannot be derived from classical equations but needs quantum mechanics for its description [33]. This situation resembles transitions in atoms: while stimulated transitions can be described within the semiclassical approach, where an atom is treated quantum mechanically and light classically, the derivation of spontaneous transitions needs quantization of light [34].

A consistent description of SPDC can be found in Ref. [33]; here we will briefly reproduce its derivation from the nonlinear optics principles. In this approach, we write the energy of interaction between light and matter, and then consider it as a Hamiltonian, which in quantum mechanics is an operator.

12.1.1 The Hamiltonian of SPDC

To generate spontaneous parametric down-conversion, it is necessary to send a coherent beam (the pump) into a crystal with second-order susceptibility. If the crystal is properly oriented, there will be photon pairs—or beams, at strong pumping—generated at the output. Here we will see why this happens; the explanation involves the Hamiltonian, i. e., the quantum operator of energy. In the dipole approximation, the energy of interaction between light and matter is, up to a sign, the product of the electric field $\vec{E}$ and the dipole moment $\vec{d}$ of the matter,

$$\mathcal{H} = -\vec{d}\vec{E}, \tag{12.1}$$

where the dipole moment is the polarization of the matter integrated over the volume,

$$\vec{d} = \int d^3 r \vec{P}(\vec{r}). \tag{12.2}$$

https://doi.org/10.1515/9783110668025-012

The integral should be done over the whole volume where the nonlinear interaction takes place, and $P(\vec{r})$ is the nonlinear polarization.

Consider first the second-order nonlinear polarization $\vec{P}^{(2)}$ (the third-order one will be responsible for four-wave mixing and similar effects in Section 12.2). The expression for $\vec{P}^{(2)}$ has been introduced in Chapter 10. After substituting Eq. (10.2) and Eq. (12.2) into Eq. (12.1), we obtain

$$\mathcal{H} = -\epsilon_0 \int d^3 r \chi^{(2)}(r) \vdots \vec{E}(\vec{r},t)\vec{E}(\vec{r},t)\vec{E}(\vec{r},t). \tag{12.3}$$

The three dots, as before, mean multiplication of a rank-3 tensor by three vectors. We assume the field $\vec{E}$ contains three components: the pump $\vec{E}_0$, the signal $\vec{E}_s$, and the idler $\vec{E}_i$. In reality, only the pump is present at the input of the nonlinear crystal; the other two fields are accounted for formally, keeping in mind that the corresponding modes are populated by only vacuum states. The reason is that in the quantum description (Chapter 11), we have to assign a field operator to every field mode, and only later, in order to calculate some observables, average the operators over states, as given by Eq. (11.50). Writing the total field $\vec{E}$ in terms of the analytic signals, we get

$$\vec{E}(\vec{r},t) = \vec{E}_0^{(+)}(\vec{r},t) + \vec{E}_s^{(+)}(\vec{r},t) + \vec{E}_i^{(+)}(\vec{r},t) + c.\,c. \tag{12.4}$$

The energy (12.3) will comprise many different terms, every one describing some nonlinear optical process. For instance, the term containing $[E_0^{(+)}]^2 E_0^{(-)}$ will correspond to the second-harmonic generation from the pump. But as we know from Chapter 10, only those processes will be efficient, for which the phase-matching conditions are satisfied. This will not be the case for all nonlinear effects. In this section, we are interested only in SPDC, and the relevant term in the expression for the energy is

$$\mathcal{H}_{SPDC} = -\epsilon_0 \int d^3 r \chi^{(2)}(r) \vdots \vec{E}_0^{(+)}(\vec{r},t)\vec{E}_s^{(-)}(\vec{r},t)\vec{E}_i^{(-)}(\vec{r},t) + c.\,c. \tag{12.5}$$

The interaction described by Eq. (12.5) takes place in the area shown in Fig. 12.1, namely where the pump beam is in the crystal of length L with the second-order nonlinear susceptibility $\chi^{(2)}$.

The pump, usually a laser beam, can be seen as a plane monochromatic classical wave propagating along the z direction, $\vec{E}_0^{(+)}(\vec{r},t) = \vec{e}_0 E_0 e^{-i\omega_0 t + ik_0 z}$, $\vec{e}_0$ being the unity

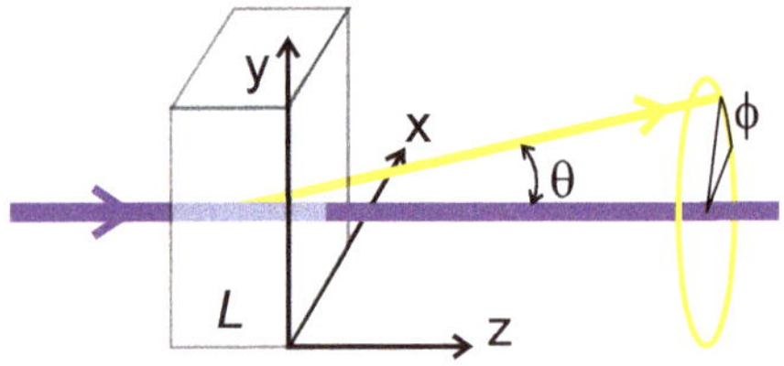

Figure 12.1: The geometry of SPDC. The pump propagates along the z axis and the nonlinear crystal has a length L.

Jones vector defining the polarization state of the pump. The situation with the signal and idler fields is different: because they are *vacuum fields*, we write $\vec{E}_s^{(-)}(\vec{r}, t)$ and $\vec{E}_i^{(-)}(\vec{r}, t)$ as quantum operators, according to Eq. (11.4):

$$\vec{E}_s^{(-)}(\vec{r}, t) = \sum_m \vec{e}_m c_m^* a_m^\dagger e^{i\omega_m t - i\vec{k}_m \vec{r}},$$

$$\vec{E}_i^{(-)}(\vec{r}, t) = \sum_n \vec{e}_n c_n^* a_n^\dagger e^{i\omega_n t - i\vec{k}_n \vec{r}}. \qquad (12.6)$$

Here, the indices m, n number all wavevector modes of the signal and idler fields, $\vec{e}_{m,n}$ are the Jones vectors defining their polarization states, and $\omega_{m,n} \equiv \omega(\vec{k}_{m,n})$ are their frequencies. We will further assume that the second-order nonlinear susceptibility has a constant value χ_0 over the whole nonlinear crystal.

Then the energy is

$$\hat{\mathcal{H}}_{\text{SPDC}} = -\epsilon_0 E_0 \sum_{m,n} c_m^* c_n^* \chi_0 \colon \vec{e}_0 \vec{e}_m \vec{e}_n \int d^3 r\, a_m^\dagger a_n^\dagger e^{i\Delta\vec{k}_{mn}\vec{r} - i\Delta\omega_{mn}t} + h.\,c., \qquad (12.7)$$

where $\Delta\omega_{mn} \equiv \omega_m + \omega_n - \omega_0$ and $\Delta\vec{k}_{mn} \equiv \vec{k}_m + \vec{k}_n - \vec{k}_0$ are frequency and wavevector mismatches, respectively. The energy (12.7) is now an operator, a *Hamiltonian*; this is why we wrote the conjugated part as 'Hermitian conjugated', 'h. c.', instead of 'complex conjugated', 'c. c.'

Every term in Hamiltonian (12.7) contains two photon creation operators, $a_m^\dagger$ and $a_n^\dagger$. This means it will lead to the generation of photon pairs in the corresponding modes, which we labeled by m, n. Due to the sum in the Hamiltonian, the pairs can be generated into many different signal and idler modes: each of the indices m, n implies three degrees of freedom in the wavevector space, as well as the polarization (see Section 11.1.1). However, there are additional restrictions related to the frequency and wavevector mismatches entering the Hamiltonian.

Consider first the frequency mismatch. If it is nonzero, the Hamiltonian will oscillate in time, and the nonlinear interaction will not be accumulated. Therefore, it is necessary that

$$\omega_m + \omega_n = \omega_0. \qquad (12.8)$$

Equation (12.8) means that the frequencies of the two generated photons should sum up to give the pump frequency. In the simplest case of $\omega_m = \omega_n$ (*frequency-degenerate SPDC*), the frequencies of the generated photons are half the pump frequency. A similar condition we obtained in Chapter 10, where the frequency of the second harmonic was twice the frequency of the pump. Now, in the 'photon language', we can say that

Eq. (12.8) formulates the energy conservation: the energy of the pump photon $\hbar\omega_0$ is equal to the sum of the daughter photon energies, $\hbar\omega_m$ and $\hbar\omega_n$.[1]

The volume integration in Eq. (12.7), $d^3r \equiv dxdydz$, imposes additional restrictions on the modes n, m, in which the signal and idler photons are generated. For simplicity, we can assume that the transverse size of the pump is very large. Then the integration over the transverse coordinates (x and y in Fig. 12.1) will lead to a factor $\delta(\Delta k_x)\delta(\Delta k_y)$ in the Hamiltonian: the transverse wavevector mismatch is equal to zero. In the 'photon language', this means conservation of the transverse momentum of photons. Because the pump photons had only momentum along the z axis, their transverse momenta were zero, and for the signal and idler photons, the projections of the momenta on both x and y axes should be opposite. The integration in z (Fig. 12.1) is over the length of the crystal, and as in all coherent nonlinear optical effects, it leads to the expression

$$\int_{-L}^{0} e^{i\Delta k_z z} = Le^{-\frac{i\Delta k_z L}{2}} \operatorname{sinc} \frac{\Delta k_z L}{2}, \tag{12.9}$$

where $\operatorname{sinc}(x) \equiv \frac{\sin x}{x}$.

The conditions that the frequency mismatch and the transverse wavevector mismatch are zero remove part of the summation in Eq. (12.7). Namely, out of the six sums over the wavevector modes, conditions $\Delta\omega = \Delta k_x = \Delta k_y = 0$ eliminate three. In addition, the Hamiltonian contains the factor (12.9), which is nonzero only in the vicinity of the longitudinal phase matching $\Delta k_z = 0$. All this leaves much less freedom in the choice of modes where the signal and idler photons are generated: unrestricted remains the frequency of one of them (let it be the signal frequency ω_s—the idler frequency is anti-correlated to it through the condition (12.8)), the azimuthal angle of emission ϕ_s of the signal photon, and the polarization states of the signal and idler photons. The latter will be the subject of the following sections.

Then the Hamiltonian can be simplified to

$$\hat{\mathcal{H}}_{SPDC} = i\hbar\Gamma a_s^\dagger a_i^\dagger + h.c., \tag{12.10}$$

where we deliberately dragged out the Planck constant and included all relevant parameters (crystal length, second-order nonlinearity, the amplitude of the pump, etc.) into the coupling parameter Γ. We also selected only one mode for the signal photon and, correspondingly, one for the idler photon, but we will keep in mind that there is still a choice of signal photon frequencies. This property of producing new states of light not into a single mode but into multiple modes, among many other features, distinguishes SPDC from the second-harmonic generation where, for instance, the spectrum of generated light is just the spectrum of the pump shifted (in the logarithmic

---

[1] This picture is valid in the case of a continuous-wave pump or a pump with relatively long pulses. For femtosecond-pulse pump, Eq. (12.8) is satisfied up to the pump spectral width; this does not mean that the energy is not conserved but only that the energy of a short pump pulse has an uncertainty.

scale) by an octave. In contrast, SPDC produces new frequency and wavevector states and, most importantly for this book, various polarization states.

12.1.2 Types of phase matching for SPDC

The polarization modes of the signal and idler photons will be determined by the type of phase matching, described in Section 10.2. In SPDC, the phase matching is usually satisfied by using a birefringent crystal and choosing different polarizations for the pump and for the signal/idler photons. For instance, in type-I SPDC in a negative crystal, the pump is polarized as an extraordinary beam while the down-converted photons are ordinarily polarized (e→oo interaction, reverse to the process discussed in Section 10.2.2). They can still differ in frequency or direction of emission, but distinguished is the case where they also have the same frequencies and wavevectors. The Hamiltonian (12.10) then takes the form

$$\hat{\mathcal{H}}_I = i\hbar\Gamma\left[a^\dagger\right]^2 + h.\,c., \tag{12.11}$$

where $a^\dagger$ is the photon creation operator in the mode into which SPDC photons are emitted. In accordance with the phase matching conditions, these photons should have the frequency equal to half of the pump frequency and the wavevector collinear with the pump wavevector. This type of SPDC phase matching is found in exactly the same way as for the second-harmonic generation (see Section 10.2.2).

The states generated through this type of phase matching manifest various remarkable quantum features, such as non-monotonic photon-number distribution, strong photon bunching, and quadrature squeezing. But because these properties are not relevant for the polarization quantum optics, we will not discuss them here.

In a more general case of type-I SPDC, the daughter photons are emitted noncollinearly with the pump (Fig. 12.2) but along cones whose opening angles depend on the frequencies of the signal and idler photons (and if they are not equal, the photon with a lower frequency will be emitted along a larger cone). The angle ϑ between the optic axis ζ and the pump wavevector defines the angle of emission for photons of a given frequency. In particular, for a certain angle, photons at the degenerate frequency $\omega_0/2$ will be emitted along the pump, and the case of collinear degenerate type-I SPDC will be realized.

One can also implement type-II phase matching, for which signal and idler photons are polarized orthogonally. The Hamiltonian then takes the form

$$\hat{\mathcal{H}}_{II} = i\hbar\Gamma a_H^\dagger a_V^\dagger + h.\,c., \tag{12.12}$$

where we assumed that the polarization states of the two photons are horizontal and vertical and the other parameters of the signal and idler photons, such as wavelength

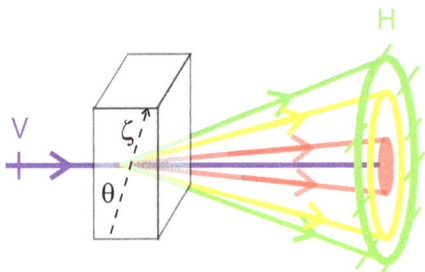

Figure 12.2: Type-I SPDC: the pump is polarized vertically (in the plane of the optic axis ζ) and the photon pairs, horizontally. Polarization directions are shown with lines.

and wavevector direction, are the same. This case is called type-II collinear frequency-degenerate SPDC. In general, the two photons are polarized as the ordinary and extraordinary normal waves in the crystal, i. e., linearly and orthogonally to each other. For simplicity, we assume that the crystal is oriented so that the directions of linear polarization are horizontal and vertical.

This case of type-II SPDC is formally similar to other cases where the signal and idler photons are emitted into two distinguishable modes. These can be different frequency modes, or different wavevector directions, or both. But in the context of this book, type-II SPDC is most interesting because it provides special polarization states, whose properties (hidden polarization, polarization squeezing) will be discussed in the next sections.

Type-II SPDC can be considered as reversed type-II second-harmonic generation (Section 10.2.2); the necessary orientation of the nonlinear crystal and the effective value of the second-order susceptibility should be calculated the same way.

Even more interesting in connection with the polarization states of nonclassical light is the phase matching involving two polarization modes and two other modes, for instance, two different wavevector directions. In this case, polarization-entangled photons are generated. This situation, depicted in Fig. 12.3, has been first realized by Kwiat et al. [38]. Importantly, SPDC produces photon pairs not only along the pump wavevector, but in a continuum of other directions and frequencies—unlike the second-harmonic generation, as already mentioned. In particular, even for a given (degenerate) frequency of the signal and idler photons $\omega_s = \omega_i = \omega_0/2$, SPDC can be non-collinear: the wavevectors of the two daughter photons are in this case not parallel to the pump wavevector. Calculation shows that, for type-II frequency-degenerate SPDC, the ordinary (o) and extraordinary (e) photons are emitted along two different

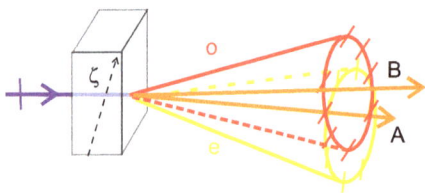

Figure 12.3: Generation of polarization-entangled states via non-collinear type-II SPDC. Lines show the polarization. Ordinary (o) and extraordinary (e) photons are emitted along red and yellow cones, respectively. The cones intersect along the directions A, B, in which entangled pairs propagate.

cones, tilted with respect to each other in the plane containing the incident pump wavevector and the optic axis (Fig. 12.3). These cones intersect along two lines, denoted as A and B in the figure.

The situation shown in Fig. 12.3 is the most general one. As the angle ϑ between the optic axis and the incident pump is varied, the cones become larger or smaller. In particular, for a certain angle ϑ they touch along a single line that is collinear with the pump wavevector. This is the case of collinear degenerate type-II SPDC we described previously.

But consider now the case shown in Fig. 12.3, and namely the photon pairs emitted in the directions A,B. Along each line, there is both an e-polarized photon (V) and an o-polarized photon (H). Because of the transverse wavevector matching condition, $\Delta k_x = \Delta k_y = 0$, the two photons should be always emitted symmetrically with respect to the pump. They also should have orthogonal polarizations. There are therefore two possibilities: that photon A is H-polarized and photon B, V polarized, and vice versa. The Hamiltonian can then be rewritten as the sum of two Hamiltonians:

$$\hat{\mathcal{H}}_{\text{ent}} = i\hbar\Gamma(a_{\text{AH}}^\dagger a_{\text{BV}}^\dagger + e^{i\phi}a_{\text{AV}}^\dagger a_{\text{BH}}^\dagger) + h.c, \tag{12.13}$$

where the phase ϕ can be different depending on the pump, signal, and idler phase delays in the nonlinear crystal.[2]

In addition to the non-collinear type-II SPDC, there is another experimental scheme to obtain Hamiltonian (12.13) [39]. In this scheme, two nonlinear crystals, cut for type-I phase matching as in Fig. 12.2, are placed one after another into a common pump beam (Fig. 12.4). One of the crystals is oriented with the optic axis ζ in the vertical plane and the other one, with the optic axis in the horizontal plane. If the pump is polarized diagonally, it has an extraordinary polarized component in each crystal and generates SPDC with the e→oo phase matching. Both crystals emit photon pairs at the degenerate frequency $\omega_0/2$ along the cone with the same opening angle, but the photons from the first crystal are polarized horizontally and the photons from the second one, vertically.

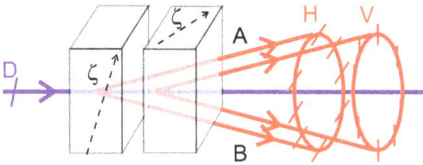

Figure 12.4: Generation of polarization-entangled states via two simultaneous non-collinear type-I SPDC processes. Lines show the polarization; the pump is polarized diagonally.

---

2 There is an additional element in the setup, not shown in Fig. 12.3: to make the two terms in Hamiltonian (12.13) coherent, a birefringent element after the nonlinear crystal has to compensate for the group-velocity delay between ordinary and extraordinary photons [38].

Let us select two directions, A and B, into which signal and idler photons are emitted. (Note that, unlike in the scheme shown in Fig. 12.3, here we can choose such directions in many different ways.) The Hamiltonian provided by the first crystal can then be written as $\hat{\mathcal{H}}_1 = i\hbar\Gamma a_{AH}^\dagger a_{BH}^\dagger + h.c.$, where $a_{AH}^\dagger$ and $a_{BH}^\dagger$ are photon creation operators in the horizontally polarized modes of beams A and B, respectively. Meanwhile, the Hamiltonian of the second crystal is $\hat{\mathcal{H}}_2 = i\hbar\Gamma a_{AV}^\dagger a_{BV}^\dagger + h.c.$, where the same notation is used.

Because the two SPDC sources are pumped coherently, by a common laser beam, the total Hamiltonian is the sum of Hamiltonians $\hat{\mathcal{H}}_1$ and $\hat{\mathcal{H}}_2$, with a constant phase between them. This phase ϕ is due to the phase delays of the pump and the emitted down-converted radiation. The result is

$$\hat{\mathcal{H}} = \hat{\mathcal{H}}_1 + e^{i\phi}\hat{\mathcal{H}}_2 = i\hbar\Gamma(a_{AH}^\dagger a_{BH}^\dagger + e^{i\phi}a_{AV}^\dagger a_{BV}^\dagger) + h.c. \tag{12.14}$$

In particular, at $\phi = 0$ the Hamiltonian becomes

$$\hat{\mathcal{H}} = i\hbar\Gamma[a_{AH}^\dagger a_{BH}^\dagger + a_{AV}^\dagger a_{BV}^\dagger] + h.c., \tag{12.15}$$

and a HWP at 45° placed in beam B will convert it into Hamiltonian (12.13) with $\phi = 0$.

This scheme is more efficient than the one based on type-II SPDC for many reasons. First, as we already mentioned, it includes many directions A and B. Second, type-I SPDC, in most crystals, has a higher effective susceptibility. Finally, this scheme is simpler in operation.

The states generated by the Hamiltonians (12.10), (12.12), (12.13) we derived here will be the subject of the next sections. But at this point we notice that each of them creates photons only in pairs. The probability of pair creation depends on the coupling parameter Γ, which scales as the second-order nonlinear susceptibility, the pump field amplitude, and the length of the crystal. Depending on the magnitude of the coupling parameter, the interaction can be weak or strong. Correspondingly, one can distinguish between two cases: low-gain SPDC, which generates photon pairs, and high-gain SPDC, which generates bright beams with photon-number correlations. We will consider the first case in Section 12.3 and the second one, in Section 12.4. But before doing that, in the next section we show how the same pair-creating Hamiltonians emerge through third-order nonlinear effects.

12.2 Spontaneous four-wave mixing and related effects

Third-order nonlinear effects occur in a broader class of materials than second-order effects because, unlike $\chi^{(2)}$, the third-order susceptibility $\chi^{(3)}$ is nonzero in practically any material. Most convenient are optical fibers, where the nonlinearity is provided by either glass or gas filling the fiber, or atomic vapors where resonances can increase the nonlinearity. Because atomic vapors, and also most of fibers, lack anisotropy, they do not offer such a rich platform for producing polarization states as crystals do. Here we will only briefly describe the methods of generating nonclassical states of polarized light in optical fibers. As in the case of SPDC, our consideration will start with deriving the Hamiltonian, i. e., the energy of the nonlinear interaction.

12.2.1 The Hamiltonian

Third-order nonlinear interaction is based on the third-order nonlinear polarization (10.3), leading to the dipole moment of the matter,

$$\vec{d} = \int d^3r \vec{P}^{(3)}(\vec{r}), \tag{12.16}$$

and the energy of the light–matter interaction,

$$\mathcal{H} = -\epsilon_0 \int d^3r \chi^{(3)}(r) \vdots \vec{E}(\vec{r}, t)\vec{E}(\vec{r}, t)\vec{E}(\vec{r}, t)\vec{E}(\vec{r}, t). \tag{12.17}$$

Similarly to the case of SPDC, we assume that the total field $\vec{E}(\vec{r}, t)$ contains the pump $\vec{E}_0(\vec{r}, t)$, which is a classical plane monochromatic wave, and the signal and idler field operators given by Eqs. (12.6). Then, among the many terms the energy will contain, we are interested in

$$\mathcal{H}_{SFWM} = -\epsilon_0 \int d^3r \chi^{(3)}(r) \vdots \vec{E}_0^{(+)}(\vec{r}, t)\vec{E}_0^{(+)}(\vec{r}, t)\vec{E}_s^{(-)}(\vec{r}, t)\vec{E}_i^{(-)}(\vec{r}, t) + c.c. \tag{12.18}$$

This term describes *spontaneous four-wave mixing* (SFWM) and *modulation instability* (MI); the difference between these two processes will be clear from what follows.

We proceed in the same way as in the case of SPDC: by substituting into Eq. (12.18) the expressions for the pump, signal and idler fields, we obtain the SFWM Hamiltonian

$$\hat{\mathcal{H}}_{SFWM} = -\epsilon_0 E_0^2 \sum_{m,n} c_m^* c_n^* \chi_0^{(3)} \vdots \vec{e}_0 \vec{e}_0 \vec{e}_m \vec{e}_n \int d^3r\, a_m^\dagger a_n^\dagger e^{i\Delta\vec{k}_{mn}\vec{r} - i\Delta\omega_{mn}t} + h.c., \tag{12.19}$$

where now, $\Delta\omega_{mn} \equiv \omega_m + \omega_n - 2\omega_0$ and $\Delta\vec{k}_{mn} \equiv \vec{k}_m + \vec{k}_n - 2\vec{k}_0$. Here, for uniformity we use the same notation (k) for the wavevectors as in the case of SPDC in nonlinear crystals. But usually, third-order interactions are implemented in optical fibers where the propagation constant β is used instead. Also, in the phase mismatch we omitted the contribution of self-phase-modulation and cross-phase-modulation; this can be done in the case of weak continuous-wave pump but would be wrong in the case of pulsed pump with a high peak power.

Equation (12.19), with the restrictions imposed by phase matching, boils down to the same pair-producing Hamiltonian (12.10) as in the case of SPDC. The difference of SFWM-MI Hamiltonians from the one of SPDC is that now, the coupling parameter scales as the third-order susceptibility and the squared pump amplitude:

$$\Gamma_{SFWM} \propto \chi^{(3)} E_0^2. \tag{12.20}$$

The fact that the coupling parameter of third-order nonlinear interactions scales quadratically with the pump amplitude leads to some important features. As will be clear from the next sections, the rate of pair production scales as the square of the coupling parameter; therefore, in SFWM and MI it will scale as the pump intensity squared. This means that, for a pulsed pump, the efficiency of SFWM and MI will be higher than for a continuous-wave pump with the same average power—the same feature is typical for the second-harmonic generation and other effects nonlinear in the pump power. Similarly, tight focusing of the pump should also increase the efficiency of SFWM and MI. In contrast, the pair generation rate of SPDC scales linearly with the pump power, and it is only the average pump power that matters.

12.2.2 Phase matching

In optical fibers, birefringence is usually absent or too small to satisfy the phase matching. Periodic poling, often used to phase match second-harmonic generation (Section 10.3.3) and SPDC, does not work here either, because poling does not change the third-order susceptibility $\chi^{(3)}$. Still, phase matching is possible, and the mechanisms are different depending on whether the pump wavelength $\lambda_0 = 2\pi c/\omega_0$ is below or above the *zero-dispersion wavelength* λ_{ZDW}.

Figure 12.5 shows a typical dispersion dependence $k(\omega)$. It is steep at low frequencies, due to the presence of infrared resonances with the molecules oscillations, and it is steep again at high frequencies, approaching electronic resonances. In between, there is the zero-dispersion frequency $\omega_{ZDW} = 2\pi c/\lambda_{ZDW}$, where the dispersion dependence has an inflection point. The dispersion dependence is convex below the zero-dispersion point and concave above it. The two intervals on the left and on the right of ω_{ZDW} are called, respectively, anomalous and normal group-velocity dispersion (GVD) ranges.

The phase-matching condition requires that the pump wavevector k_0 is the mean arithmetic of the signal and idler wavevectors, $2k_0 = k_s + k_i$, while the pump frequency ω_0 is the mean arithmetic of the signal and idler frequencies, $2\omega_0 = \omega_s + \omega_i$. If the pump frequency is above the zero-dispersion frequency ω_{ZDW}, as shown by the blue dashed line in Fig. 12.5, it is possible to satisfy this condition. This is geometrically illustrated by the blue solid line in the figure, connecting points s1 and i1 on the dispersion dependence. The phase-matching condition can be satisfied in the normal GVD range, as long as the pump (p1) is not too far from the zero-dispersion point. This regime of pair generation is called spontaneous four-wave mixing, and it produces signal and idler

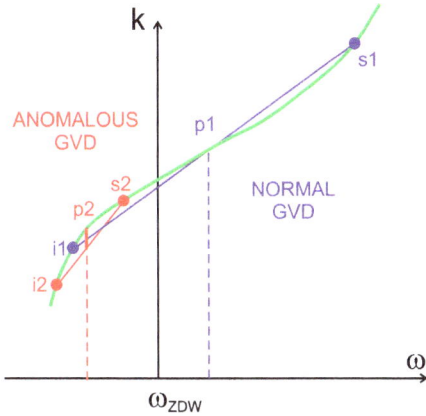

Figure 12.5: Two regimes of pair generation through third-order nonlinear interaction: four-wave mixing (blue) and modulation instability (red). Dashed lines show pump frequencies and dots denote signal (s) and idler (i) pairs.

photons at relatively large separated frequencies. These frequencies are solely determined by the dispersion of the nonlinear material and do not depend on the pump power [52].

But if the pump frequency (shown by a red dashed line) is below ω_{ZDW}, i. e., in the anomalous GVD range, the pump wavevector cannot be the mean arithmetic of the signal and idler wavevectors. This is shown in Fig. 12.5 by a red straight line connecting two points s2, i2 on the dispersion dependence that are symmetric with respect to ω_0. The mean arithmetic of $k(\omega_s)$ and $k(\omega_i)$ is always below the dispersion dependence. And here the cross- and self-phase-modulation come into play. We ignored them so far, because these effects change the phase matching very little, but it still matters if only a small wavevector mismatch has to be compensated. Indeed, if the pump is strong, it changes both its refractive index (self-phase modulation) and the refractive indices of the signal and idler radiation (cross-phase-modulation) [2]. These two effects add to the *negative* mismatch $\Delta k = k_s + k_i - 2k_0$ a *positive* term scaling as the pump power, which reduces the absolute value of the mismatch. This term, shown by red vertical bar in Fig. 12.5, can be viewed as reducing the wavevector of the pump and therefore making the phase matching satisfied. Naturally, the stronger the pump, the larger the nonlinear change in the refractive index, and hence the further apart the signal and idler wavelengths. This regime is called modulation instability. In this regime, signal and idler photons are generated spectrally rather close to the pump, but their frequency separation considerably increases with the increase in the pump power [52].

In the regime of moderately pumped MI, where the signal and idler photons are generated even closer to the pump frequency (as so-called sidebands of the pump), the emergence of nonclassical light has a different interpretation [49]. In this case, the pump and the signal/idler sidebands can be considered as a single strong beam. Due to the Kerr effect, the refractive index of this beam gets a nonlinear additional part, caused by self-phase-modulation. This additional part scales linearly with the intensity I of light,

$$n = n_0 + n_2 I, \tag{12.21}$$

where n_2 is called the nonlinear refractive index. If a coherent state enters a fiber, its intensity is constant only in the classical picture; from the quantum-optical viewpoint even a coherent state has shot-noise fluctuations of the amplitude (Section 11.1.2) and photon-number/intensity fluctuations. Its Wigner function (Section 11.1.3) is Gaussian, with the width of 1/2 in each quadrature. This Wigner function is schematically shown in Fig. 12.6 with a circle. Without the Kerr nonlinearity the state would only change its phase, i. e., the Wigner function would simply rotate in the phase space with the circular frequency ω_0. Now we will look at the Wigner function as at a real probability distribution of the quadratures, i. e., as a set of points in the phase space.[3] Due the nonlinear change of the refractive index, points with larger amplitudes acquire larger phase shifts than points with smaller amplitudes (dashed lines in the figure). After a sufficiently long nonlinear medium (fiber), the Wigner-function distribution stretches, as shown in Fig. 12.6 with an ellipse. Note, however, that the Kerr effect does not change the amplitude of light, only its phase; therefore the stretching occurs along a certain quadrature q_a that is different from the amplitude. The amplitude uncertainty remains the same, and the anti-squeezing of quadrature q_a leads to the squeezing of the orthogonal quadrature q_s.

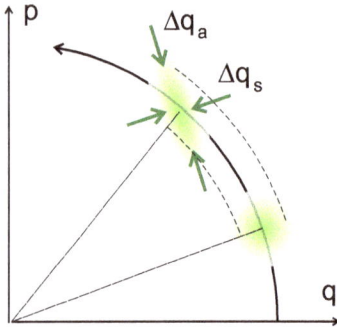

Figure 12.6: Quadrature squeezing resulting from third-order (Kerr-type) nonlinear interaction. While the photon-number uncertainty remains at the shot-noise level, squeezing appears in the quadrature q_s at the expense of anti-squeezing in quadrature q_a.

Phase matching is satisfied in this 'Kerr squeezing' effect automatically, because the signal and idler sidebands have only minute spectral separation from the pump. What is important, however, is that, for very short pulses, the interaction is reduced by the GVD of the fiber. To mitigate this problem, it is useful to work in the soliton regime of pulse propagation, where the spreading of a pulse is compensated by the Kerr effect [4, 49].

---

3 This is not a rigorous picture, just a classical interpretation (see Section 11.1.3), but it gives a good intuition on how the Kerr nonlinearity leads to squeezing.

12.2.3 Engineering more complicated interactions

Third-order nonlinear interactions discussed in the previous two sections all lead to the pair-generating Hamiltonians of the form (12.10). The difference between SFWM, MI, and Kerr squeezing is only that in the first case, the frequencies ω_s, ω_i of the signal and idler photons are well separated from the pump frequency, in the second case, very little separated, and in the last case, not separated at all. This does not leave much freedom for engineering the Hamiltonian and the resulting quantum states, especially from the polarization viewpoint. The signal and idler photons are typically emitted into the same polarization mode as the pump.

SFWM also enables the interaction where both photons are emitted into the same mode, i. e., where the Hamiltonian has the form (12.11). Then, for the signal/idler mode not to coincide with the one of the strong pump, two pump beams are used instead of one. These pump beams are separated in frequency or, in the case of SFWM in atoms, in wavevector.

Of course, if the signal and idler photons have frequencies sufficiently far apart, their polarization states can be transformed independently and, for instance, orthogonally polarized pairs can be produced, like in the case of Hamiltonian (12.12). But the signal and idler photons will still differ in frequency, which is a restriction in some cases.

The solution to this engineering problem is similar to the one where two SPDC crystals are used. This time (Fig. 12.7(a)), two fibers are placed one after another into the same pump beam (green in the figure), each of them generating signal and idler photons in sidebands A, B (shown with blue and yellow colors) through some third-order process, for instance, SFWM. If the pump is polarized vertically, signal and idler photons at the output of the first fiber are also polarized vertically. Further, a HWP oriented at 45° rotates the polarization of signal and idler photons by 90° but does not change the pump polarization. Then the second fiber will also generate signal and idler photons polarized vertically. At the output of the second fiber, there will be vertically polarized pairs from the second fiber, but also horizontally polarized pairs from the first fiber. The total Hamiltonian will have the form (12.15). The only difference from the SPDC case shown in Fig. 12.4 is that there beams A and B have different directions, and now they have the same direction but differ in frequencies. Alternatively, the HWP can only change the pump polarization and maintain the polarization states of the signal and idler photons; then the pairs from the first fiber will be polarized vertically and from the second fiber, horizontally.

This strategy is part of a more general principle of 'nonlinear interferometry': if two nonlinear processes are pumped coherently, there is interference between them. Nonlinear effects within the same polarization mode can enhance or suppress each other, but fields emitted into orthogonal polarization modes can form new superpositions and therefore new polarization states [12].

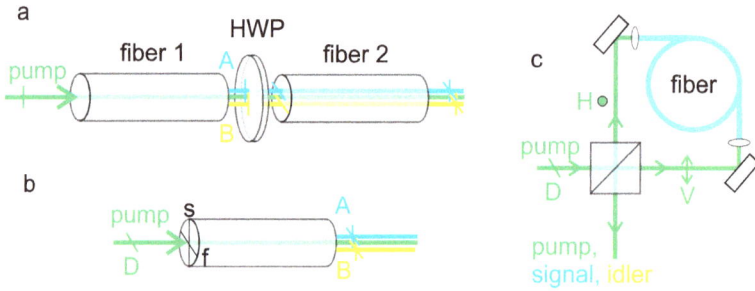

Figure 12.7: Interferometric schemes to generate photon pairs in optical fibers: two fibers placed into a common pump beam (a), the use of fast and slow axes of a polarization-maintaining fiber (b), and the Sagnac interferometer (c).

Usually, selective polarization rotation for only signal and idler, or only for the pump, is difficult—and impossible if signal and idler frequencies coincide with the pump one. To solve this problem in the case of Kerr squeezing, Heersink et al. [20] used, instead of two different fibers, a single polarization-maintaining fiber (Section 9.5.1). Because such a fiber suppresses the cross-talk between fields polarized along the 'fast' and 'slow' axes, nonlinear interaction occurs for both polarizations independently. With the pump polarized diagonally (Fig. 12.7(b)) and slow (s) and fast (f) axes corresponding to V and H polarizations, the fiber produced both vertically and horizontally polarized photon pairs (Fig. 12.7(b)). The group delay between such pairs had to be pre-compensated [20].

Another solution is to use an interferometer, for instance, a Sagnac interferometer shown in Fig. 12.7(c). In a very elegant way, this strategy was applied to SFWM [50]. A diagonally polarized pump was split at a polarizing beamsplitter, and the horizontally and vertically polarized beams propagated in the Sagnac loop clockwise and anticlockwise, respectively. Correspondingly, signal and idler photons from horizontally polarized pump, also horizontally polarized, were reflected into the output port of the Sagnac interferometer, and overlapped with the vertically polarized pairs, which were transmitted. Again, Hamiltonian (12.15) was realized. An unbalanced Sagnac interferometer was earlier applied to Kerr squeezing [45, 48], in a more complicated scheme that we will not discuss here, but later replaced by the same group with a scheme based on a polarization-maintaining fiber (Fig. 12.7(b)).

12.3 Low parametric gain and entangled photons

After having described various schemes to implement the Hamiltonians, we pass to the description of the quantum states generated. Let us first address the case where the nonlinear interaction is weak. Typically, this happens when the pump is a continuous-wave laser with moderate power (tens of milliwatts). In this case, it is convenient to find the quantum state by using the Schrödinger approach in combination with the perturbation theory.

12.3.1 Perturbation theory: photon pairs

In the Schrödinger approach, the state $|\Psi(t)\rangle$ varies with the time due to the action of the Hamiltonian, and the operators are considered to be time-independent. The initial state of the modes where signal and idler photons are generated is the vacuum state, because there are no signal or idler photons at the input of the crystal. The evolution of the state is then described by the Schrödinger equation with one of the Hamiltonians $\hat{\mathcal{H}}$ we derived in the previous sections:

$$i\hbar\frac{d|\Psi(t)\rangle}{dt} = \hat{\mathcal{H}}|\Psi(t)\rangle. \tag{12.22}$$

The solution is then

$$|\Psi(t)\rangle = e^{\frac{1}{i\hbar}\int \hat{\mathcal{H}}dt}|\Psi(0)\rangle, \tag{12.23}$$

where the initial state $|\Psi(0)\rangle$ is the vacuum. Because the Hamiltonians we derived in Section 12.1 are time-independent, the integration in Eq. (12.23) boils down to the multiplication by the integration time (time t_i of the interaction). Then the magnitude of the exponent in Eq. (12.23) is given by Γt_i. This parameter is the key characteristic of the 'strength' of pair-producing interaction, SPDC or SFWM, and will be further denoted as $G = \Gamma t_i$ and called the parametric gain.

If the interaction is weak, $G \ll 1$, the exponential in Eq. (12.23) can be expanded into a Taylor series, with only the first two terms kept, which yields

$$|\Psi\rangle = |0\rangle + G a_s^\dagger a_i^\dagger |0\rangle, \tag{12.24}$$

where $|0\rangle$ is the vacuum states for all involved modes: a single mode in the case of degenerate collinear SPDC, polarization modes H and V for collinear type-II SPDC, and four modes AH, AV, BH, BV for the processes shown in Figs. 12.3, 12.4, and 12.7.

The state (12.24) is a superposition of the vacuum state and a state formed by action of two photon creation operators on the vacuum state. Depending on whether the photon creation operators belong to the same mode or different modes, the second term in Eq. (12.24) describes either a two-photon Fock state or a product of single-photon Fock states in two orthogonal polarization modes. If the Hamiltonian involves photon creation operators in four modes, as in Eq. (12.13), the resulting state is polarization-entangled and will be considered in Section 12.3.3. But in all cases the state (12.24) describes a superposition of the vacuum and a photon pair, a *biphoton*.

12.3.2 Orthogonally polarized photon pairs and polarization Hong–Ou–Mandel effect

Even in the simplest case where SPDC produces photon pairs into the same mode, the state at the output of the nonlinear crystal,

$$|\Psi\rangle = |0\rangle + G[a^\dagger]^2|0\rangle = |0\rangle + \sqrt{2}G|2\rangle, \tag{12.25}$$

is nonclassical. For instance, its bunching parameter $g^{(2)} = 1/(2|G|^2)$ is very high at $|G| \ll 1$, and condition (11.36) for $n = 2$ is satisfied even if higher-order terms in the expansion (12.25) are taken into account and calculation gives a nonzero third-order correlation function $g^{(3)}$. This *extreme bunching* can be observed by sending the state into a Hanbury Brown–Twiss setup (Fig. 11.2): the vacuum part of the state will not affect the experiment, but the photon pairs, with a probability 50 %, will be split on the beamsplitter, and the detectors will 'click' simultaneously in each such case. If the photon flux is low, accidental coincidences will be very few, much fewer than those caused by photon pairs, and the resulting bunching parameter will be high.

Condition (11.37) is also satisfied for state (12.25), for even m. To see this, we can continue the Taylor expansion in Eq. (12.25) and obtain the result that it contains no three-photon Fock state but only a four-photon one, and so on. The state produced by frequency-degenerate collinear type-I SPDC is for this reason called sometimes 'light with even photon numbers' [30]. But from the viewpoint of polarization properties, this state of light is not very interesting, and we will pass now to the state generated via frequency-degenerate collinear type-II SPDC.

This state has the form

$$|\Psi\rangle = |0\rangle + Ga_H^\dagger a_V^\dagger|0\rangle = |0\rangle + G|1\rangle_H|1\rangle_V. \tag{12.26}$$

It also consists of biphotons, but now the photons within one biphoton are orthogonally polarized. Their correlation can be measured in a modified Hanbury Brown–Twiss setup, where a non-polarizing beamsplitter is replaced by a polarizing one (Fig. 12.8). The two detectors will *always* 'click' simultaneously in this case, as long as there are no losses. This way, one can measure the second-order cross-correlation function (11.26) for the H and V polarization modes, and the result will be $g_{H,V}^{(2)}(0) = 1/|G|^2 \gg 1$.

This type of biphotons has a very important application: by using one of the detectors in Fig. 12.8 as a trigger, one can produce single-photon states in the other channel. For instance, if the detector registering a vertically polarized photon 'clicks', we know for sure that there is a horizontally polarized photon in the 'transmitted' path of the polarization prism, and we can further use it. For instance, we can open a gate in the transmitted arm if there is a photon in the reflected arm. The probability that a second pair is accompanying the first one is very low if the parametric gain is low; therefore only a single photon will pass through the gate, and the state in the transmitted arm

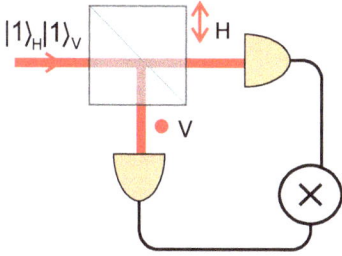

Figure 12.8: A setup for observing photon correlations in type-II SPDC.

will be very close to the Fock state $|1\rangle$. This method is called *heralded preparation of single photons*, and it is used, for instance, for quantum key distribution. It does not necessarily require type-II SPDC: it is only necessary that photon pairs are emitted into two distinguishable modes. For instance, in the pioneering experiment by Hong and Mandel [21], the two photons were emitted in two different directions. But the polarization version is technically simpler.

A very unusual feature of both the state (12.26) and its two-photon part $|1\rangle_H|1\rangle_V$ is that despite being pure, it is completely unpolarized. Indeed, calculation of the Stokes parameters for this state yields $\langle S_1 \rangle = \langle S_2 \rangle = \langle S_3 \rangle = 0$ and $\langle S_0 \rangle = 2|G|^2$. This looks strange at first sight, but we need to recall that two orthogonally polarized photons (or, generally, electric fields) will interfere and give a polarized state only if they are coherent. Meanwhile, the two photons generated via SPDC are not coherent with each other—although they are correlated. Figure 12.9 illustrates this feature: it shows the pair of orthogonally polarized photons $|1\rangle_H|1\rangle_V$ on the Poincaré sphere. If there is no coherence between the two photons, their Stokes vectors should be summed geometrically and yield zero.[4] Therefore the degree of polarization of state (12.26) is zero. In Section 12.5 we will see that this state and other similar states feature what is called 'hidden polarization', and how this behavior suggests alternative definitions for the degree of polarization (Section 12.5.3).

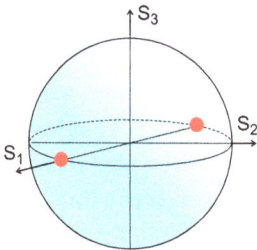

Figure 12.9: The state $|1\rangle_H|1\rangle_V$ represented by two points on the Poncaré sphere.

---

4 Using two SPDC sources in a configuration as in Fig. 12.4, together with polarization transformations at the input and output, one can prepare biphotons in a single frequency and wavevector mode with an arbitrary degree of polarization [9].

A question arises whether the state $|1\rangle_H|1\rangle_V$ is an entangled state. As written here, it is a factorable state because it is a product of two single-photon Fock states in two orthogonal polarization modes. Indeed, if the quantum states are understood as states populating certain modes of radiation, the state $|1\rangle_H|1\rangle_V$ is perfectly factorable. There is, however, another attitude [15]: if the two photons of the state are 'labeled', say, as 's' and 'i' (signal and idler), then the two-photon state should be 'symmetrized' with respect to the photon exchange and written as

$$|\Psi\rangle = \frac{1}{\sqrt{2}}(|H\rangle_s|V\rangle_i + |V\rangle_s|H\rangle_i). \tag{12.27}$$

Equation (12.27) means that either the signal photon is horizontally polarized and the idler photon, vertically polarized, or vice versa.[5] This type of entanglement is similar to another debatable case, namely of the entanglement of a single photon in two arms of a Mach–Zehnder interferometer [35] or a single photon in two polarization modes [36].

Regardless of whether to consider the state (12.26) as entangled or factorable, it can be projected on an entangled state [6]. Indeed, let us split it on a non-polarizing beamsplitter, and label as 's' and 'i' the photons in different arms. If we ignore the pairs that are directed into the same arm (transmitted or reflected) of the beamsplitter, the rest of the state will be written as in Eq. (12.27) and be entangled, because it is not factorable into some states of photons 's' and 'i', or modes 's' and 'i'.

The state (12.26) manifests one of the most intriguing effects in quantum optics, namely the Hong–Ou–Mandel dip; however, in its polarization version. In the 'standard' version of the Hong–Ou–Mandel effect [22], two photons are overlapped on a 50 % beamsplitter (Fig. 12.10(a)). If the photons are indistinguishable in all parameters, namely frequency, wavevector direction, polarization, and perfectly overlapped in time and space, then they are never directed into two different output ports of the beamsplitter. They are always in the same output port, and a pair of detectors placed in different output ports will never 'click' simultaneously. If the time delay t between the arrivals of the two photons on the beamsplitter is larger than the coherence time of the photons, they 'do not notice each other' and are independently split on the beamsplitter; the detectors then 'click' in coincidence with the probability 50 %, i. e., very often. This forms a 'dip' in the rate R_c of coincidence counting.

In the polarization version of the Hong–Ou–Mandel effect [44, 46, 47] (Fig. 12.10(b)), the two-photon part of state (12.26) is transformed by a HWP oriented at 22.5° into $|1\rangle_D|1\rangle_A$, a pair of diagonally and anti-diagonally polarized photons, and then sent to a polarizing beamsplitter. The beamsplitter is oriented so that it transmits a horizontally polarized photon and reflects a vertically polarized photon. Then each

---

5 Actually in quantum mechanics a superposition state means not 'either–or' but 'both'—similar to how a single photon in the Young experiment can pass, in principle, through two slits at the same time and interfere with itself.

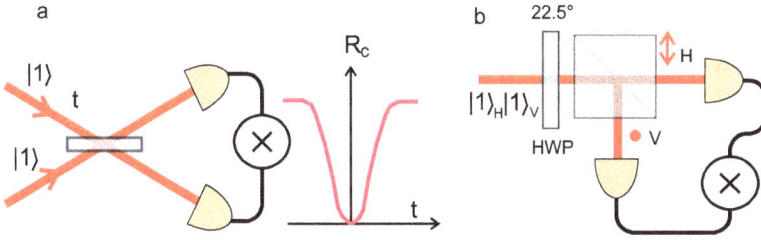

Figure 12.10: The 'standard' Hong–Ou–Mandel effect (a) and its polarization version (b).

of the photons of the $|1\rangle_D|1\rangle_A$ pair has a 50 % chance to be reflected or transmitted. But due to the interference, the pair always goes into the same port, and there are no coincidences between the photocounts of two detectors after the polarizing beamsplitter in Fig. 12.10(b). We will now derive this result rigorously, highlighting the similarity with the 'standard' HOM effect.

As mentioned in Chapter 3, a polarization transformation is similar to a two-mode beamsplitter transformation, and it is described by a similar Jones matrix. In particular, a HWP (Chapter 5) placed at an angle 22.5° performs the same transformation on polarization modes as a 50 % beamsplitter on spatial modes, and its Jones matrix is (see Eq. (5.11))

$$J_{HWP} = \frac{i}{\sqrt{2}} \begin{pmatrix} 1 & 1 \\ 1 & -1 \end{pmatrix}. \tag{12.28}$$

Let us find the state after the HWP. We will do it by writing the input state as $a_H^\dagger a_V^\dagger |0\rangle$, then transforming the photon annihilation operators a_H and a_V with the Jones matrix (12.28), and Hermitian conjugating them to obtain photon creation operators. For the new operators in the H, V modes we get

$$a_H'^\dagger = -\frac{i}{\sqrt{2}}(a_H^\dagger + a_V^\dagger),$$

$$a_V'^\dagger = -\frac{i}{\sqrt{2}}(a_H^\dagger - a_V^\dagger). \tag{12.29}$$

Inverting this transformation to obtain $a_H^\dagger$, $a_V^\dagger$ in terms of $a_H'^\dagger$, $a_V'^\dagger$ and substituting these expressions into the two-photon part of the state (12.26), after the HWP we get a biphoton of the form

$$|\Psi\rangle^{(2)} = \frac{1}{2}\left([a_V'^\dagger]^2 - [a_H'^\dagger]^2\right) = \frac{1}{\sqrt{2}}(|2\rangle_V - |2\rangle_H). \tag{12.30}$$

(The superscript '(2)' denotes the two-photon part of the state.)

We see that after the plate, the horizontal and vertical polarization modes are always occupied by two-photon Fock states, and there are no photon pairs of the form $|1\rangle_H|1\rangle_V$. The state (12.30) will not produce any coincidences at the output of the setup shown in Fig. 12.10(b). Of course, this effect is only possible if the input horizontally and vertically polarized photons are indistinguishable; in reality, there is always a time delay between the orthogonally polarized photons at the output of a type-II SPDC process. This delay is caused by the difference between the group velocities of the ordinary and extraordinary waves in the nonlinear crystal and can be compensated in experiment by birefringent plates inserted after it [47].

12.3.3 Polarization-entangled photons and Bell states

Consider now the most complicated Hamiltonian realized through SPDC, SFWM, or MI, namely the one of Eq. (12.13), corresponding to the situation shown in Figs. 12.3, 12.4, and 12.7. As everywhere in Section 12.3, here we assume that the parametric gain, equal to the product of Γ and the interaction time t_i, is small. Then the generated state will be, as in all other cases of this section, a superposition of the vacuum and a two-photon state. Its two-photon part, for instance, in the case of Fig. 12.3, is

$$|\Psi\rangle^{(2)} = \frac{1}{\sqrt{2}}(|H\rangle_A|V\rangle_B + e^{i\phi}|V\rangle_A|H\rangle_B). \tag{12.31}$$

Here we simplified the notation by writing a single-photon Fock state in mode AH as $|H\rangle_A$ etc. The sign in this superposition depends on the phase shift in the crystal. The state (12.31) is a polarization-entangled state because it cannot be represented as a product of a certain polarization state in mode A and another state in mode B.

In particular, the states with $\phi = 0$ and $\phi = \pi$ are denoted as

$$|\Psi^{(+)}\rangle \equiv \frac{1}{\sqrt{2}}(|H\rangle_A|V\rangle_B + |V\rangle_A|H\rangle_B), \tag{12.32}$$

$$|\Psi^{(-)}\rangle \equiv \frac{1}{\sqrt{2}}(|H\rangle_A|V\rangle_B - |V\rangle_A|H\rangle_B). \tag{12.33}$$

The two states (12.32) and (12.33) are polarization-entangled biphotons with photons in a pair polarized orthogonally.

These states can be easily converted into biphotons where both photons in a pair are in the same polarization states. This can be done by placing a HWP oriented at 45° into arm B, thus rotating the polarization of photon B by 90°. This gives the states

$$|\Phi^{(+)}\rangle \equiv \frac{1}{\sqrt{2}}(|H\rangle_A|H\rangle_B + |V\rangle_A|V\rangle_B), \tag{12.34}$$

$$|\Phi^{(-)}\rangle \equiv \frac{1}{\sqrt{2}}(|H\rangle_A|H\rangle_B - |V\rangle_A|V\rangle_B). \tag{12.35}$$

Equations (12.32)–(12.35) describe the so-called *Bell states*, the most notorious example of entangled photons. They are named so due to their role in the tests of Bell's inequalities, which will be the subject of Chapter 13. The states (12.32)–(12.35) are maximally polarization-entangled and orthogonal to each other.

One of the distinguishing features of an entangled state of two subsystems A, B is that taken separately, each subsystem is in a mixed state [43]. The polarization analog of a mixed state is an unpolarized state. Accordingly, for each of the photons A, B the degree of polarization is zero. This can be verified by calculating the Stokes parameters for photons A and B in the Bell states. In experiment, if a Stokes measurement setup (Fig. 11.4(b)) is placed in each path, A and B in Fig. 12.3, at any settings of the HWP and QWP the upper and lower detectors will 'click' equally frequent. However, there will be correlation between their 'clicks'. For instance, if the state at the input is $|\Phi^{(-)}\rangle$ and all phase plates are oriented at $0°$ (the first Stokes observable is measured), the upper and lower detectors in paths A and B will always register photons simultaneously: if photon A is detected with the horizontal polarization, its match photon B also has horizontal polarization.

Due to the symmetry with respect to the exchange of the photons, the state $|\Psi^{(-)}\rangle$ is called the *singlet state*, while the other three Bell states $|\Phi^{(+)}\rangle$, $|\Phi^{(-)}\rangle$, and $|\Psi^{(+)}\rangle$ are said to form a *triplet*. The singlet state has a remarkable property: it maintains its form in any polarization basis [17, 41]. For instance, in the diagonal and circular bases it remains a pair of orthogonally polarized photons:

$$|\Psi^{(-)}\rangle = \frac{1}{\sqrt{2}}(|D\rangle_A|A\rangle_B - |A\rangle_A|D\rangle_B) = \frac{1}{\sqrt{2}}(|R\rangle_A|L\rangle_B - |L\rangle_A|R\rangle_B). \tag{12.36}$$

The other Bell states can be obtained from this one by means of local (i. e., in only one beam) or global (in both beams) polarization transformations.

12.4 High parametric gain: polarization squeezing and entanglement

So far, we only considered low-gain SPDC, SFWM, and MI, and entangled photons that these processes generate. What happens when the nonlinear interaction becomes strong? Apparently, the flux of photon pairs gets so large that different pairs overlap in time, and the simultaneity of photon arrival at two detectors is masked by accidental overlaps of different pairs. Indeed, this is what one observes, for instance, in the Hanbury Brown–Twiss setup (Fig. 11.2): the number of accidental coincidences grows quadratically with the photon flux, the number of real coincidences grows linearly, and if the photon flux is high, the former hugely exceeds the latter. Then the bunching parameter $g^{(2)}(0)$, as well as the cross-correlation function for signal and idler modes, gets very close to unity. Nevertheless, the output radiation remains nonclassical in this case. To analyze its properties, in this section we will use the *Heisenberg picture* [34], in which the state is considered to be constant but the operators undergo evolution.

12.4.1 Evolution of operators

Instead of finding the quantum state at the output of the nonlinear source, here we will look at the operators, such as photon creation, annihilation, and photon-number operators in both polarization modes. In the end, we are only interested in observable quantities [32], and these are mean photon numbers or various statistical moments (variances, correlation functions etc.) For this purpose, the Heisenberg picture is at least as suitable as the Schrödinger one; moreover, it has the advantage of treating operators similar to fields in the classical optics; it is therefore more intuitive [34].

The time evolution of an operator $\hat{A}$ is governed by the Heisenberg equation

$$i\hbar \frac{d\hat{A}}{dt} = [\hat{A}, \hat{\mathcal{H}}], \tag{12.37}$$

where $\hat{\mathcal{H}}$ is the Hamiltonian.

For a Hamiltonian of the form (12.11), the Heisenberg equation is

$$\frac{da}{dt} = 2\Gamma a^{\dagger}, \tag{12.38}$$

and the equation for $a^{\dagger}$ is obtained by Hermitian conjugation. The solution for the operator a is [33, 53]

$$a(t) = \cosh(2\Gamma t)a(0) + \sinh(2\Gamma t)a^{\dagger}(0), \tag{12.39}$$

where $a(0)$ is the initial operator, i. e., the operator before the evolution imposed by the Hamiltonian. As in the previous section, we define the interaction time t_i and denote $\Gamma t_i \equiv G$, the parametric gain. Equation (12.39) is called the Bogolyubov transformation.

For the quadrature operators (11.9) $\hat{q}$, $\hat{p}$, the Bogolyubov transformation leads to [53]

$$\hat{q}(t) = e^{2G}\hat{q}(0), \quad \hat{p}(t) = e^{-2G}\hat{p}(0). \tag{12.40}$$

These transformations mean that the quadratures evolve along hyperbolas $q(t)p(t) = $ const, which is shown in Fig. 12.11 [11]. As in Section 12.2.2, we use the Wigner-function

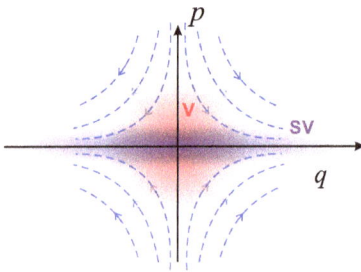

Figure 12.11: The evolution in the phase space due to Hamiltonian (12.11) and the emergence of squeezed vacuum (blue) from the vacuum state (red).

representation; now we assume that initially, the state is a vacuum, represented by an uncertainty region (group of points) at the origin. Due to the action of the Hamiltonian (12.11) each point evolves along a hyperbola. The resulting state will be stretched along one direction and squeezed along the other direction (q and p, respectively, if $G > 0$). This state is called squeezed vacuum: vacuum, because it is not displaced from the origin in the phase space, and squeezed, because its uncertainty in one quadrature is smaller than the uncertainty in the other one. Indeed, it follows from Eq. (12.40) that the variances of the quadratures will evolve as

$$\text{Var}[q(t)] = e^{4G}\,\text{Var}[q(0)], \quad \text{Var}[p(t)] = e^{-4G}\,\text{Var}[p(0)]. \tag{12.41}$$

Because the initial state is the vacuum, $\text{Var}[q(0)] = \text{Var}[p(0)] = 1/4$, the final state shows quadrature squeezing:

$$\Delta p(t) = \frac{e^{-2G}}{2} < \frac{1}{2} < \Delta q(t) = \frac{e^{2G}}{2}. \tag{12.42}$$

From the Bogolyubov transformation (12.39), we can find the mean photon number in the squeezed vacuum state. It is found by averaging the output photon-number operator over the input (vacuum) state:

$$\langle \hat{N} \rangle \equiv \langle 0|a^\dagger(t)a(t)|0 \rangle$$
$$= \langle 0|(\cosh(2G)a^\dagger(0) + \sinh(2G)a(0))(\cosh(2G)a(0) + \sinh(2G)a^\dagger(0))|0 \rangle. \tag{12.43}$$

The initial (vacuum) photon annihilation operator $a(0)$ yields zero after acting on the vacuum; accounting for this and using commutation relations in (12.43), we obtain

$$\langle \hat{N} \rangle = \sinh^2(2G). \tag{12.44}$$

We see that at the output of SPDC, the mean photon number can be very large if the parametric gain G is high. For instance, in experiments with strongly pumped SPDC (usually the pump is pulsed), values of G as high as 8 can be obtained, leading to mean numbers of photons on the order of 10^{13} [11]. This regime of SPDC is known as *high-gain parametric down-conversion (PDC)*, and the state at the output is called *bright squeezed vacuum*. The term 'bright' here means that the number of photons per radiation mode is high—as high as in laser radiation [11].

We will now consider the Hamiltonians that involve both polarization modes: the type-II SPDC Hamiltonian (12.12) and the 'entangling' Hamiltonian (12.13).

In the case of type-II SPDC, the Heisenberg equations for the annihilation operators in both polarization modes are

$$\frac{da_H}{dt} = \Gamma a_V^\dagger,$$
$$\frac{da_V}{dt} = \Gamma a_H^\dagger. \tag{12.45}$$

The solution to these equations, as one can verify by substitution, is given by the two-mode Bogolyubov transformations,

$$a_H(t) = \cosh(\Gamma t)a_H(0) + \sinh(\Gamma t)a_V^\dagger(0),$$
$$a_V(t) = \cosh(\Gamma t)a_V(0) + \sinh(\Gamma t)a_H^\dagger(0). \tag{12.46}$$

Here, $a_H(0)$ and $a_V(0)$ are initial operators, i. e., operators before the evolution imposed by the SPDC Hamiltonian. As in the previous case, we define the interaction time t_i and denote $\Gamma t_i \equiv G$, the parametric gain.

From the Bogolyubov transformations (12.46), we can find various parameters of the SPDC radiation, similar to the case of Hamiltonian (12.11).

The mean numbers of photons in the horizontal and vertical polarization modes are found as in the previous case, and are

$$\langle N_H \rangle = \langle N_V \rangle = \sinh^2 G. \tag{12.47}$$

We obtain the result that the mean photon numbers in the horizontal and vertical polarization modes are the same. This is not surprising because the Hamiltonian (12.11) was symmetric with respect to the interchange of these modes. Moreover, Eq. (12.47) shows that these photon numbers can be very large—the parametric gain is twice as small as in the case of degenerate collinear high-gain PDC, but still can be very high. But the most remarkable feature follows from the fact that type-II SPDC generates photons always in pairs, so that every time a photon appears in mode H, its twin appears in mode V. This leads to the effect of polarization squeezing, which will be the subject of the next section.

12.4.2 Polarization squeezing

From $\langle N_H \rangle = \langle N_V \rangle$, it follows that the mean value of the first Stokes operator is zero, $\langle S_1 \rangle = 0$. By calculating the other Stokes parameters, one can also verify that $\langle S_2 \rangle = \langle S_3 \rangle = 0$. But the most surprising and, in fact, nonclassical feature of the radiation produced through high-gain type-II PDC is that the variance of the first Stokes observable is zero as well. Indeed, fluctuations of the photon numbers in modes H and V are perfectly correlated, because photons appear in these two modes simultaneously. Therefore, the difference of these photon numbers, i. e., the first Stokes observable S_1, does not fluctuate.

According to quantum mechanics, the first Stokes observable does not fluctuate simply because the operator $\hat{S}_1$ commutes with Hamiltonian (12.12):

$$[\hat{S}_1, \hat{\mathcal{H}}_{II}] = [a_H^\dagger a_H - a_V^\dagger a_V, i\hbar a_H^\dagger a_V^\dagger + h.c.] = 0. \tag{12.48}$$

Therefore, the first Stokes observable has no fluctuations and, in particular,

$$\mathrm{Var}(\hat{S}_1) = 0. \tag{12.49}$$

This perfectly noiseless behavior can be only observed in the absence of losses and under the condition of perfect detection efficiency. In reality this is not the case. If the detection efficiency, including all losses on the way from the generation to the detection, is η for both H and V polarization modes, then the variance of the first Stokes observable is [11]

$$\mathrm{Var}(\hat{S}_1) = (1 - \eta)\langle \hat{S}_0 \rangle. \tag{12.50}$$

But even with non-unity detection efficiency, the variance of the first Stokes observable is smaller than $\langle S_0 \rangle$, i.e., the total number of photons. Meanwhile, for a coherent state the variances of all Stokes observables are equal to $\langle S_0 \rangle$, which can be considered as the shot-noise limit for polarization noise. This effect, when fluctuations in one of the Stokes observables are reduced below the shot-noise limit,

$$\mathrm{Var}(\hat{S}_i) < \langle \hat{S}_0 \rangle, \tag{12.51}$$

has been defined as polarization squeezing [13, 31]. Any Stokes observable can be squeezed, from $i = 1, 2, 3$ to any generic Stokes observable (11.42). Obviously, by an appropriate polarization transformation the squeezing can be transferred from the first Stokes observable to any generic one.

Polarization squeezing is a special case of twin-beam squeezing (Section 11.1.3), where the variance of the photon-number difference between two beams is less than the mean total number of photons in both beams. As any twin-beam squeezing, polarization squeezing is a nonclassical feature.

Polarization squeezing has been first reported by Bushev et al. [10] who observed it at the output of a type-II parametric oscillator operating below threshold. In this regime, a parametric oscillator emits bright squeezed vacuum, similar to high-gain PDC. The Hamiltonian realized in this case was (12.12), and squeezing was in the first Stokes observable. Polarization tomography revealed a polarization quasi-probability (Section 11.3) whose width in S_2 and S_3 directions exceeded the shot noise but whose width in the S_1 direction was below the shot noise. This polarization quasi-probability distribution is schematically shown in Fig. 12.12 (red shape, labeled 1). It is at the center of the Stokes space because, as mentioned above, Hamiltonian (12.12) generates unpolarized light, $\langle S_1 \rangle = \langle S_2 \rangle = \langle S_3 \rangle = 0$. For comparison, the green shape 2 shows in this figure the polarization quasi-probability of a coherent circularly polarized state with the same mean number of photons.

Suppression of noise in polarization observables is useful in all measurements based on polarization, an obvious example being polarimetry. Like in all measurements, suppression of the noise below the shot-noise level offers a quantum advantage

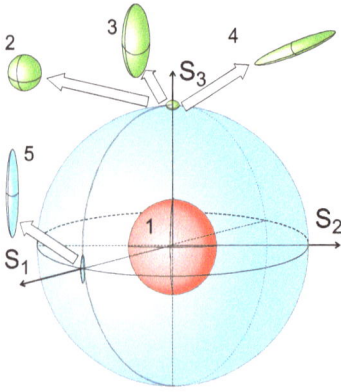

Figure 12.12: Schematic view of polarization quasi-probabilities for states with polarization squeezing generated by Bushev et al. [10] (1), by Bowen et al. [7] (3, 4), and by Heersink et al. [19] (5). State 2 is a circularly polarized coherent state.

over classical measurements, which in the best case use coherent light. For instance, Feng and Pfister [16] achieved signal-to-noise ratio in polarimetry 4.8 dB better than with coherent states. Using polarization squeezing in an atomic magnetometer, Wolfgramm et al. [54] demonstrated an advantage of 3.2 dB over a coherent-state measurement.

Polarization squeezing is related to quadrature squeezing. For instance, by passing from H, V to A, D modes, we can rewrite Hamiltonian (12.12) in the form

$$\hat{\mathcal{H}}_{II} = \frac{i\hbar\Gamma}{2}\left([a_D^\dagger]^2 - [a_A^\dagger]^2\right) + h.\,c., \tag{12.52}$$

which is the difference of two single-mode Hamiltonians. Here, we see exactly the same effect as in the experiment of Fig. 12.4: two pairs of orthogonally polarized photons are equivalent to a coherent superposition of a photon pair in the same polarization state and another photon pair in the orthogonal polarization state. The same interference phenomenon underpins the polarization Hong–Ou–Mandel effect (Fig. 12.10(b)). In other words, one can obtain polarization squeezing in the first Stokes variable S_1 ('pancake-like' shape 1 in Fig. 12.12) by placing two type-I SPDC crystals one by one into a common pump beam, the first crystal producing diagonally (D) polarized pairs and the second crystal, anti-diagonally (A) polarized pairs. Each of these crystals will then produce quadrature-squeezed vacuum; correspondingly, the two terms in Hamiltonian (12.52) describe quadrature squeezing in modes A, D.

Bowen et al. [7] used this principle to produce polarization squeezing by combining two quadrature-squeezed beams polarized orthogonally. In their experiment, beams were polarized horizontally (H) and vertically (V); clearly, the resulting state was not squeezed in S_1. Depending on the phase between the two beams, squeezing was obtained either in S_2 (green 'pancake-like' shape 3 in Fig. 12.12) or in both S_2 and S_3 (green 'cigar-like' shape 4 in Fig. 12.12). But because the initial quadrature-squeezed beams were not squeezed vacuums but displaced squeezed states, with coherent components, the resulting states also had a coherent component. This component was polarized, as would always be the case for a superposition of two coherent orthogonally polarized states. The states therefore were displaced in the Stokes space: they had $\langle \hat{S}_3 \rangle = \langle \hat{S}_0 \rangle$ and $\langle \hat{S}_1 \rangle = \langle \hat{S}_2 \rangle = 0$, i. e., they were circularly polarized (Fig. 12.12).

Remarkably, while the 'pancake-like' state produced in Ref. [7] was not squeezed in any Stokes observable except S_2, the 'cigar-like' state was additionally squeezed in S_0, i. e., it had sub-shot-noise intensity fluctuations. Moreover, the polarization squeezing of this state was of a different nature than the one of the squeezed vacuum state obtained by Bushev et al. [10]. Indeed, for state 4 in Fig. 12.12 not only the noise in some Stokes observables is less than for a coherent state with the same mean photon number (shape 2 in Fig. 12.12). Additionally, this state satisfies the inequality

$$\text{Var}(\hat{S}_1) < |\langle \hat{S}_3 \rangle| < \text{Var}(\hat{S}_2), \tag{12.53}$$

i. e., its polarization squeezing has a meaning with respect to the uncertainty relations (11.47) for the Stokes observables. In contrast, a polarization-squeezed vacuum state is unpolarized, and the right-hand sides of uncertainty relations (11.47) are zero for this state.

Partly for this reason, and also because the condition (12.51) of polarization squeezing boils down to the quadrature squeezing in two polarization modes A, D, Korolkova et al. [19, 36, 37] proposed an alternative definition for the polarization squeezing. In this definition, a state of light is polarization-squeezed if

$$\text{Var}(\hat{S}_i) < |\langle \hat{S}_j \rangle| < \text{Var}(\hat{S}_k), \quad i \neq j \neq k. \tag{12.54}$$

This type of polarization squeezing was also obtained by overlapping two orthogonally polarized Kerr-squeezed pulses, using a Sagnac interferometer (Fig. 12.7(c)) [19] and later, two polarization modes in a polarization-maintaining fiber (Fig. 12.7(b)). The state produced by Heersink et al. [19] is shown as blue shape 5 in Fig. 12.12. For such a state, Marquardt et al. [42] performed the polarization tomography and observed a 'cigar-like' polarization quasi-probability distribution.

12.4.3 Polarization entanglement

Let us now return to the experimental schemes involving two polarization modes and two other modes, for instance, wavevector (directional), as in Figs. 12.3, 12.4. The corresponding Hamiltonians have the form (12.13) or (12.14). At low parametric gain, the states at the output are pairs of polarization-entangled photons. At high parametric gain, as in the cases considered in previous sections, the flux of pairs becomes very strong. What are the properties and, in particular, nonclassical features of the output states? As always in the case of bright light, it is convenient to treat this problem in the Heisenberg picture. We will use H and V polarization modes and also the modes A and B, which can be wavevector (directional) or frequency modes. We will only consider Hamiltonian (12.13) with $\phi = 0$, which at low parametric gain produces Bell state $|\Psi^{(+)}\rangle$, but we will keep in mind that there are other three 'entangling' Hamiltonians, producing the other three Bell states at low gain.

The Heisenberg equations for four operators a_{AH}, a_{AV}, a_{BH}, a_{BV} lead to the four-mode Bogolyubov transformations

$$
\begin{aligned}
a_{AH}(t) &= \cosh(\Gamma t)a_{AH}(0) + \sinh(\Gamma t)a_{BV}^\dagger(0), \\
a_{AV}(t) &= \cosh(\Gamma t)a_{AV}(0) + \sinh(\Gamma t)a_{BH}^\dagger(0), \\
a_{BH}(t) &= \cosh(\Gamma t)a_{BH}(0) + \sinh(\Gamma t)a_{AV}^\dagger(0), \\
a_{BV}(t) &= \cosh(\Gamma t)a_{BV}(0) + \sinh(\Gamma t)a_{AH}^\dagger(0).
\end{aligned}
\tag{12.55}
$$

We can notice that these four equations form two independent pairs: two equations (the first and the fourth) involve creation and annihilation operators in modes AH, BV, and the other two (the second and the third) involve operators in modes BH, AV. In both cases, the relations between two operators are the same as between a_H, a_V in the Bogolyubov transformations (12.46). We therefore expect that the operator $\hat{N}_{AH} - \hat{N}_{BV}$, i. e., the difference of photon numbers in modes AH, BV will not fluctuate, or at least, even in the presence of loss will have fluctuations below the shot-noise limit. The same will be the case for the operator $\hat{N}_{BH} - \hat{N}_{AV}$. Summing both operators, we see that *the difference of the total photon numbers in polarization modes H and V will not fluctuate*. We can come to the same conclusion by noticing that the operator $\hat{N}_{AH} + \hat{N}_{BH} - \hat{N}_{AV} - \hat{N}_{BV}$ commutes with the Hamiltonian. This leads to several important properties of the state.

In 1993, Karassiov and Masalov [27, 28] proposed to consider the resulting state in terms of joint Stokes observables for modes A and B, introduced as

$$
\hat{S}_i^{\text{joint}} \equiv \hat{S}_i^A + \hat{S}_i^B, \quad i = 0, 1, 2, 3.
\tag{12.56}
$$

These joint Stokes observables have a physical meaning if A and B are wavelength modes. Then observables (12.56) can be measured in a standard Stokes measurement setup as shown in Fig. 5.7 or, in the quantum case, in Fig. 11.4(b). The setup should not distinguish between modes A and B and the detectors should measure the total number of photons.

It follows that, for the state described by Hamiltonian (12.13) with $\phi = 0$, the first joint Stokes observable will have, ideally, no noise—and noise reduced below the shot-noise limit even under imperfect detection and losses. The polarization probability distribution for such a state should be 'pancake-like', ideally, infinitely thin along S_1 and having a large size (noise considerably larger than the shot noise) along the other two observables. This probability distribution is shown by blue color in Fig. 12.13. Its polarization properties are similar to the ones of the Bell state $|\Psi^{(+)}\rangle$ but, unlike a two-photon entangled state, it contains a large (macroscopic) number of photons. This state can be called a *macroscopic Bell state* $|\Psi_{mac}^{(+)}\rangle$ [24]. It was generated in a below-threshold optical parametric oscillator by Bushev et al. [10], and later using a system of two type-I SPDC crystals (as in Fig. 12.4) pumped by strong picosecond pulses by Iskhakov et al. [24]. Its polarization tomography [10] indeed showed a 'pancake-like' shape as in Fig. 12.13.

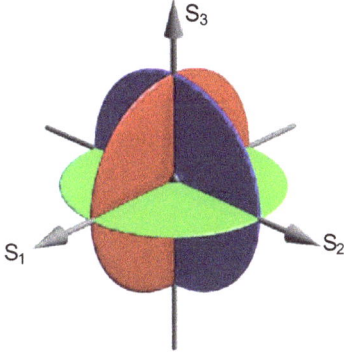

Figure 12.13: Macroscopic Bell states in the space of Stokes variables: $|\Psi_{mac}^{(+)}\rangle$ (blue), $|\Phi_{mac}^{(-)}\rangle$ (red), $|\Phi_{mac}^{(+)}\rangle$ (green), and $|\Psi_{mac}^{(-)}\rangle$ (black point at the center).

Similarly, by stronger pumping a system that at low parametric gain generates the other three Bell states, (12.34), (12.35), and (12.33), one can obtain macroscopic Bell states $|\Phi_{mac}^{(+)}\rangle$, $|\Phi_{mac}^{(-)}\rangle$, and $|\Psi_{mac}^{(-)}\rangle$. Their polarization quasi-probability distributions are shown in Fig. 12.13 with green, red, and black colors, respectively. Each of the states $|\Phi_{mac}^{(+)}\rangle$ and $|\Phi_{mac}^{(-)}\rangle$ is squeezed in one Stokes observable: the first one in S_3, the second one in S_2.

Most interesting is the *macroscopic singlet Bell state* $|\Psi_{mac}^{(-)}\rangle$, shown in Fig. 12.13 as a black point at the origin. Because, similarly to the two-photon singlet Bell state, it is invariant to polarization transformations, and because, similarly to the state $|\Psi_{mac}^{(+)}\rangle$, it should feature squeezing of the first Stokes observable, *it should have squeezed fluctuations of all Stokes observables*. This is why this state is shown as a point at the center of the Stokes space. Because of these unusual properties, theoretically predicted by Karassiov and Masalov [27], it was called '*polarization-scalar light*'. This state was generated by Iskhakov et al. [25], and its polarization tomography [26] showed that indeed, the fluctuations of all Stokes observables in this state were suppressed below the shot-noise level.

Another remarkable property of the macroscopic singlet Bell state is that it violates inequality (11.49) for the joint Stokes observables. Indeed, it was experimentally shown in Ref. [23] that, for this state,

$$\text{Var}(\hat{S}_1) + \text{Var}(\hat{S}_2) + \text{Var}(\hat{S}_3) < 2\langle \hat{S}_0 \rangle. \tag{12.57}$$

In the absence of loss, inequality (12.57) turns into

$$\text{Var}(\hat{S}_1) + \text{Var}(\hat{S}_2) + \text{Var}(\hat{S}_3) = 0. \tag{12.58}$$

The contradiction with inequality (11.49) is because the latter was derived for a state of light in a single frequency and wavevector mode; meanwhile, the singlet state and all other macroscopic Bell states generated in Refs. [10, 23, 24, 26] contained at least two frequency modes (A,B). Nevertheless, Eq. (12.58) leads to a remarkable result [23]. In the state $|\Psi_{mac}^{(-)}\rangle$, photons are emitted into modes A and B in large groups, of 10^5 on the average. One can place Stokes measurement setups (Fig. 5.7) in each mode A and B, and count photons that are reflected and transmitted by polarization prisms in each case. If the settings of waveplates in beams A and B are the same, the number n of photons reflected in arm A is uncertain, but always the same as the number of photons transmitted in arm B (Fig. 12.14). The same is true for the number m of transmitted photons in arm A: it will be always the same as the number of reflected photons in arm B.

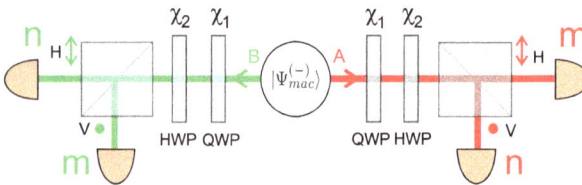

Figure 12.14: Correlations in photon numbers observed for the macroscopic singlet Bell state.

This result means that, for the macroscopic singlet Bell state, each Stokes observable for each mode A, B is uncertain, but there are correlations in the Stokes observables between modes A and B. This property, according to the general definition, can be considered as entanglement. Indeed, as shown in ref. [23], it can be interpreted as polarization entanglement for macroscopic pulses of light.

However, it was pointed out by Korolkova et al. [36, 37] that an unpolarized state like $|\Psi_{mac}^{(-)}\rangle$ or any polarization-squeezed vacuum state is irrelevant for the Stokes observables uncertainty relations, because for such a state $\langle S_i \rangle = 0$, $i = 1, 2, 3$, and the right-hand sides of uncertainty relations (11.47) are zero. Therefore, despite the correlation between the quantum Stokes observables for modes A,B in Fig. 12.14, such a situation does not enable simultaneous accurate measurement of two non-commuting Stokes observables for the same mode, with an 'apparent violation of the uncertainty principle' [36]. The definition of polarization entanglement was formulated in Ref. [36] as the condition

$$\text{Var}(\hat{S}_i^A | S_i^B) \, \text{Var}(\hat{S}_j^A | S_j^B) < |\langle \hat{S}_k \rangle|^2, \quad i \neq j \neq k, \tag{12.59}$$

where $\text{Var}(\hat{S}_i^A | S_i^B)$ is the variance of the ith Stokes observable for mode A conditioned on the measurement of the same Stokes observable for mode B.

Polarization entanglement of the form (12.59) has been reported by Bowen et al. [8] using a modification of their experiment on polarization squeezing [7] and by Dong et al. [14] using Kerr squeezing in optical fibers.

12.5 'Hidden polarization'

Chapter 12, at least according to the title, reviews the nonclassical states of polarized light. Meanwhile, its several sections deal with the states that are unpolarized if we accept definition (3.30) for the degree of polarization: the mean values of all Stokes observables are zero for these states. This section shows how most of these states turn out to be polarized if not only the mean values of the Stokes observables are considered, but also higher-order moments. This effect, predicted by Klyshko [29], is known as 'hidden polarization'.

12.5.1 'Hidden polarization' of nonclassical light

The most basic definition (3.30) of the degree of polarization dictates the following way of measuring it [31]: in a Stokes measurement setup (Fig. 5.7), the orientations of the HWP and QWP are varied in all possible ways, and the visibility in the intensity modulation measured by one of the detectors gives the degree of polarization [Chapter 5, Eq. (5.26)]. According to this definition, orthogonally polarized photon pairs (Section 12.3.2), polarization-entangled photon pairs including the Bell states (Section 12.3.3), and the macroscopic analogues of all these states emitted through high-gain PDC (Section 12.4) are unpolarized. But at the same time, for most of these states, rotation of the waveplates in the scheme of Fig. 5.7 does lead to some observable modulation. This modulation, however, is not in the mean intensity but in higher-order intensity moments.

Examples have been already given in several sections of this Chapter. Consider, for instance, the polarization Hong–Ou–Mandel effect (Section 12.3.2, Fig. 12.10(b)). If the HWP is at 22.5°, the two detectors are almost never 'clicking' in coincidence. But if the HWP is oriented at 0°, the detectors will always 'click' in coincidence. As the HWP is rotated, the rate of single counts will be constant for each detector, but the rate of coincidence counts will be 100 % modulated. 'Distinguished' directions, in which maximum coincidence rate will be measured, will be the ones of the initial polarizations of the photons in pairs (H,V). This is a typical example of hidden polarization. The result of such a measurement is shown in Fig. 12.15 with a red solid line. It is sufficient to remove the QWP and rotate only the HWP (left panel in the figure); as its orientation χ_2 is varied, the rate of coincidences is modulated. The maxima correspond to the cases where the photons are not split on the beamsplitter: the horizontally polarized photon goes to one detector and its vertically polarized 'match', to the other one. As mentioned, the count rates of both detectors do not show any modulation as the HWP is rotated.

Two-photon Bell states (Section 12.3.3) manifest a similar behavior. For instance, in the scheme of Fig. 12.4, the radiation in any direction is fully unpolarized, according to definition (3.30) or (5.26). But correlated measurement of the Stokes observables in

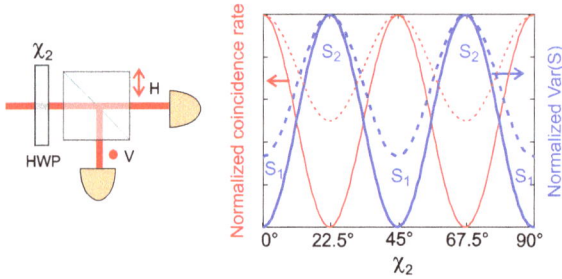

Figure 12.15: A setup to observe 'hidden polarization' (left) and the dependence of normalized coincidence rate (red thin solid line) and normalized variance of the Stokes observable (blue thick solid line) on the HWP orientation (right) in the case of SPDC radiation at the input. The mean count rates of both detectors (not shown) have no modulation. Dashed lines show the dependences for classical radiation at the input.

modes A and B simultaneously will reveal 'distinguished' directions for the three Bell states $|\Phi^{(\pm)}\rangle$, $|\Psi^{(+)}\rangle$. However, this example involves two radiation modes (A, B).

If the flux of photon pairs at the input of the measurement setup of Fig. 12.10(b) is very high, almost all coincidences will become accidental, as mentioned in the beginning of Section 12.4. The modulation in coincidence count rate will disappear, but it does not mean that the hidden polarization effect disappears as well. It just means that the measurement should be different. In order to observe 'hidden polarization' in the radiation of strongly pumped type-II SPDC, one should measure not the rate of coincidences but the variance of the Stokes observable that is set by the orientations of the plates [31]. For instance, as shown theoretically in Section 12.4.2, the first Stokes observable has, in the absence of losses, zero variance. Meanwhile, the variances of the S_2 and S_3 Stokes observables for this state exceed the shot-noise limit. Therefore, by measuring the variance of the difference of signals from the two detectors in Fig. 5.7 and rotating the waveplates, one will see, ideally, a 100 % modulation. Again, to see this effect in high-gain type-II PDC it suffices to have only the HWP (Fig. 12.15). Rotation of this plate leads to the modulation of the measured Stokes observable variance (blue thick solid line). The minima correspond to the measurement of S_1, for which the noise is suppressed; this happens when the HWP is at 0°, 45°, and 90°. In the positions of the HWP at 22.5° and 67.5°, the setup probes observable S_2, which has enhanced noise. At the same time, no modulation is observed in the signals of both detectors.

This behavior has indeed been reported for the radiation at the output of a type-II optical parametric oscillator under threshold by Bushev et al. [10]. The dependence of the Stokes observable noise resembled the one shown by blue solid line in Fig. 12.15, albeit with lower visibility: because of the losses and imperfect detection, the noise was never exactly zero.

For photon pairs emitted via type-II low-gain SPDC, hidden polarization was observed by Usachev et al. [51] by using the same setup and measuring coincidence count rate. Rotating a HWP in front of a polarizing prism, they obtained a dependence similar to the solid red curve in Fig. 12.15.

Experiments with the macroscopic Bell state $|\Psi^{(-)}_{mac}\rangle$ showed that this state does not have 'hidden polarization' [25]. Because this state is invariant to polarization rotation, all moments of its Stokes observables $S_{1,2,3}$ are zero. This is why this state was termed 'polarization-scalar light' [27]: it is unpolarized in all orders in the intensity.

12.5.2 Classical 'hidden polarization'

One might think that 'hidden polarization' is a typically quantum effect. This is not true: as proposed by Klyshko [31], two overlapped orthogonally polarized laser beams would produce the same effects as for SPDC radiation at the input, albeit with lower visibilities. In the case of coincidence counting, the visibility will be only 33 % (red thin dashed line in Fig. 12.15). In the case of Stokes variance measurement, 50 % visibility should be observed (blue thick dashed line in Fig. 12.15). In this measurement, it might seem that there is no difference between the quantum and classical cases, because the quantum case will give finite visibility as well due to losses and imperfect detection. But the boundary between the classical and quantum cases is set by the shot-noise limit, which can be overcome for nonclassical light.

An experimental realization of this proposal has a certain difficulty. Namely, the two lasers whose beams should be combined to obtain the 'hidden polarization' effect have to be incoherent but still of the same wavelength. As a solution, one could take a single laser beam, polarized at 45°, split it on a polarizing beamsplitter, delay one polarization component with respect to the other by more than a coherence length, and recombine the beams on a polarizing beamsplitter (Fig. 12.16). The output beam will contain an incoherent mixture of horizontally and vertically polarized components. If it is sent into a measurement setup as shown in Fig. 12.15, it will manifest 'hidden polarization'. for instance, if variances of different Stokes observables are measured, the one of S_1 will be smaller than the one of S_2.

Such an experiment has been made by Guzun and Penin [18] by registering coincidences in the scheme shown in Fig. 12.15 with an attenuated laser beam at the input.

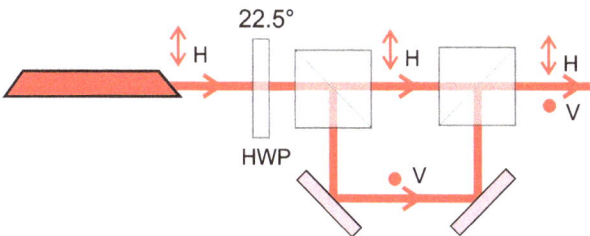

Figure 12.16: A simplified setup to impart 'hidden polarization' to a classical coherent beam.

As expected, they observed a dependence similar to the one shown in Fig. 12.15 by red thin dashed line.

12.5.3 Alternative definitions of the degree of polarization

The existence of 'hidden polarization' shows that the degree of polarization, as defined by Eq. (3.30), is incomplete: it classifies states like polarization-squeezed vacuum, or orthogonally polarized photon pairs, as unpolarized. An example has been pointed out by Agarwal and Puri [1]: in the course of propagation through optical fiber, a coherent beam can generate a quantum state through Kerr squeezing, and its degree of polarization may change. Besides, there are other features that make this definition inconvenient. For instance, a vacuum state turns out to be fully polarized [3], which is unphysical. In the framework of nano-optics (Chapter 8), definition (3.30) also has to be modified, because it takes into account only transverse (two-dimensional) polarization.

A straightforward way to generalize the degree of polarization is to replace classical averages in Eq. (3.30) by quantum averages. Similarly, in the coherence matrix (3.31) the field correlators should be replaced by quantum correlators, for instance $\langle E_H^* E_V \rangle \rightarrow \langle a_H^\dagger a_V \rangle$ [1]. However, this does not help to circumvent the problems mentioned above.

Alodjants et al. [3] introduced the degree of polarization as

$$P_A \equiv \frac{\sqrt{\langle S_1 \rangle^2 + \langle S_2 \rangle^2 + \langle S_3 \rangle^2}}{\sqrt{\langle S_1^2 \rangle + \langle S_2^2 \rangle + \langle S_3^2 \rangle}}. \tag{12.60}$$

This definition provides a zero degree of polarization for a vacuum state.

To account for the 'hidden polarization' effect, Klyshko [31] suggested to use, instead of the coherence matrix (3.31), its higher-order analogues, containing higher-order normally-ordered moments. For instance, the matrix describing the distribution of coincidence counting rates contains fourth-order moments of creation and annihilation operators, of the form $\langle [a_H^\dagger]^2 a_V^2 \rangle$, $\langle a_H^\dagger a_V^\dagger a_H a_V \rangle$, etc. Then the higher-order degree of polarization is defined, instead of (3.30), in terms of higher-order moments of the intensity. This definition has simple operational meaning: it corresponds to the visibility in the modulation of the corresponding moments measured in the setup of Fig. 12.15 under all possible settings of the two plates.

According to this definition, both the pair of orthogonally polarized photons $|1\rangle_H |1\rangle_V$ (Fig. 12.9) and a polarization-squeezed vacuum (shape 1 in Fig. 12.12) are fully polarized in the second order in the intensity. To verify this in experiment, in the first case it is more convenient to measure the coincidence count rate; in the second case, the variance of the Stokes observables. Theoretically, 100 % modulation visibility is expected in both cases; but the variance measurement is affected by the detection efficiency.

Several theoretical definitions for the degree of polarization were introduced as distances on the Poincaré sphere from a completely unpolarized state [5]. The latter is defined as a state 'spread' over the Poincaré sphere, so that it is invariant to any polarization transformation. However, these measures are not operational: they do not enable direct measurement in experiment.

Finally, to take into account the three-dimensional structure of polarization at the nanoscale (Chapter 8), as well as the vacuum fluctuations, Luis [40] proposed to generalize both the Stokes observables and the degree of polarization to include a third dimension.

Bibliography

[1]   G. S. Agarwal and R. R. Puri. Quantum theory of propagation of elliptically polarized light through a Kerr medium. *Phys. Rev. A*, 40:5179–5186, Nov 1989.

[2]   G. Agrawal. *Nonlinear fiber optics*. Academic Press, 1989.

[3]   A. P. Alodjants, S. M. Arakelian, and A. S. Chirkin. Polarization quantum states of light in nonlinear distributed feedback systems; quantum nondemolition measurements of the Stokes parameters of light and atomic angular momentum. *Appl. Phys. B, Lasers Opt.*, 66(1):53–65, January 1998.

[4]   H.-A. Bachor and T. C. Ralph. *A guide to experiments in quantum optics*. Wiley-VCH, 2004.

[5]   G. Björk, J. Söderholm, L. L. Sánchez-Soto, A. B. Klimov, I. Ghiu, P. Marian, and T. A. Marian. Quantum degrees of polarization. *Opt. Commun.*, 283:4440–4447, 2010.

[6]   D. Bouwmeester, A. Ekert, and A. Zeilinger. *The physics of quantum information*. Springer-Verlag, 2000.

[7]   W. P. Bowen, R. Schnabel, H. A. Bachor, and P. K. Lam. Polarization squeezing of continuous variable Stokes parameters. *Phys. Rev. Lett.*, 88:093601, Feb 2002.

[8]   W. P. Bowen, N. Treps, R. Schnabel, and P. K. Lam. Experimental demonstration of continuous variable polarization entanglement. *Phys. Rev. Lett.*, 89:253601, Dec 2002.

[9]   A. V. Burlakov and M. V. Chekhova. Polarization optics of biphotons. *JETP Lett.*, 75:432–438, 2002.

[10]   P. A. Bushev, V. P. Karassiov, A. V. Masalov, and A. A. Putilin. Biphoton light with hidden polarization and its polarization tomography. *Opt. Spectrosc.*, 91:558–564, 2001.

[11]   M. V. Chekhova, G. Leuchs, and M. Zukowski. Bright squeezed vacuum: entanglement of macroscopic light beams. *Opt. Commun.*, 337:27, 2014.

[12]   M. V. Chekhova and Z. Y. Ou. Nonlinear interferometers in quantum optics. *Adv. Opt. Photonics*, 8(1):104–155, Mar 2016.

[13]   A. S. Chirkin, A. A. Orlov, and D. Y. Parashchuk. Quantum theory of two-mode interactions in optically anisotropic media with cubic nonlinearities: Generation of quadrature- and polarization-squeezed light. *Rus. Journ. Quantum Electronics*, 23:870–874, 1993.

[14]   R. Dong, J. Heersink, J.-I. Yoshikawa, O. Glöckl, U. L. Andersen, and G. Leuchs. An efficient source of continuous variable polarization entanglement. *New J. Phys.*, 9(11):410, nov 2007.

[15]   M. V. Fedorov and N. I. Miklin. Schmidt modes and entanglement. *Contemp. Phys.*, 2014.

[16]   S. Feng and O. Pfister. Sub-shot-noise heterodyne polarimetry. *Opt. Lett.*, 29(23):2800–2802, Dec 2004.

[17] C. C. Gerry and P. L. Knight. *Introductory quantum optics*. Cambridge University Press, 2005.

[18] D. I. Guzun and A. N. Penin. Hidden polarization of two-mode coherent light. In S. N. Bagayev and A. S. Chirkin, editors, *Atomic and Quantum Optics: High-Precision Measurements*, volume 2799, pages 249–254. International Society for Optics and Photonics, SPIE, 1996.

[19] J. Heersink, T. Gaber, S. Lorenz, O. Gloeckl, N. Korolkova, and G. Leuchs. Polarization squeezing of intense pulses with a fiber-optic Sagnac interferometer. *Phys. Rev. A*, 68:013815, 2003.

[20] J. Heersink, V. Josse, G. Leuchs, and U. Andersen. Efficient polarization squeezing in optical fibers. *Opt. Lett.*, 30:1192, 2005.

[21] C. K. Hong and L. Mandel. Experimental realization of a localized one-photon state. *Phys. Rev. Lett.*, 56:58–60, Jan 1986.

[22] C. K. Hong, Z. Y. Ou, and L. Mandel. Measurement of subpicosecond time intervals between two photons by interference. *Phys. Rev. Lett.*, 59:2044–2046, Nov 1987.

[23] T. S. Iskhakov, I. N. Agafonov, M. V. Chekhova, and G. Leuchs. Polarization-entangled light pulses of 10^5 photons. *Phys. Rev. Lett.*, 109:150502, Oct 2012.

[24] T. S. Iskhakov, I. N. Agafonov, M. V. Chekhova, G. O. Rytikov, and G. Leuchs. Polarization properties of macroscopic Bell states. *Phys. Rev. A*, 84:045804, Oct 2011.

[25] T. S. Iskhakov, M. V. Chekhova, G. O. Rytikov, and G. Leuchs. Macroscopic pure state of light free of polarization noise. *Phys. Rev. Lett.*, 106:113602, Mar 2011.

[26] B. Kanseri, T. Iskhakov, I. Agafonov, M. Chekhova, and G. Leuchs. Three-dimensional quantum polarization tomography of macroscopic Bell states. *Phys. Rev. A*, 85:022126, Feb 2012.

[27] V. P. Karasev and A. V. Masalov. Unpolarized light states in quantum optics. *Opt. Spectrosc.*, 74:551, 1994.

[28] V. P. Karassiov. Polarization structure of quantum light fields: a new insight: I. General outlook. *J. Phys. A*, 26:4345, 1993.

[29] D. N. Klyshko. Multiphoton light and polarization effects. *Phys. Lett. A*, 163:349, 1992.

[30] D. N. Klyshko. The nonclassical light. *Phys. Usp.*, 39:573–596, 1996.

[31] D. N. Klyshko. Polarization of light: fourth-order effects and polarization-squeezed states. *J. Exp. Theor. Phys.*, 84:1065–1079, 1997.

[32] D. N. Klyshko. Basic quantum mechanical concepts from the operational viewpoint. *Phys. Usp.*, 41(9):885–922, 1998.

[33] D. Klyshko. *Photons and nonlinear optics*. Gordon and Breach, 1988.

[34] D. Klyshko. *Physical foundations of quantum electronics*. World Scientific, 2011.

[35] N. Korolkova and G. Leuchs. Quantum correlations in separable multi-mode states and in classically entangled light. *Rep. Prog. Phys.*, 8:056001, 2019.

[36] N. Korolkova, G. Leuchs, R. Loudon, T. C. Ralph, and C. Silberhorn. Polarization squeezing and continuous-variable polarization entanglement. *Phys. Rev. A*, 65:052306, Apr 2002.

[37] N. Korolkova and R. Loudon. Nonseparability and squeezing of continuous polarization variables. *Phys. Rev. Lett.*, 71:032343, 2005.

[38] P. G. Kwiat, K. Mattle, H. Weinfurter, A. Zeilinger, A. V. Sergienko, and Y. Shih. New high-intensity source of polarization-entangled photon pairs. *Phys. Rev. Lett.*, 75:4337–4341, Dec 1995.

[39] P. G. Kwiat, E. Waks, A. G. White, I. Appelbaum, and P. H. Eberhard. Ultrabright source of polarization-entangled photons. *Phys. Rev. A*, 60:R773–R776, Aug 1999.

[40] A. Luis. Quantum polarization for three-dimensional fields via Stokes operators. *Phys. Rev. A*, 71:023810, Feb 2005.

[41] A. I. Lvovsky. *Quantum physics: an introduction based on photons*. Springer, 2018.

[42] C. Marquardt, J. Heersink, R. Dong, M. V. Chekhova, A. B. Klimov, L. L. Sánchez-Soto, U. L. Andersen, and G. Leuchs. Quantum reconstruction of an intense polarization squeezed optical state. *Phys. Rev. Lett.*, 99:220401, Nov 2007.

[43] M. A. Nielsen and I. L. Chuang. *Quantum computation and quantum information*. Cambridge University Press, 2010.

[44] M. H. Rubin, D. N. Klyshko, Y. H. Shih, and A. V. Sergienko. Theory of two-photon entanglement in type-ii optical parametric down-conversion. *Phys. Rev. A*, 50:5122–5133, Dec 1994.

[45] S. Schmitt, J. Ficker, M. Wolff, F. König, A. Sizmann, and G. Leuchs. Photon-number squeezed solitons from an asymmetric fiber-optic Sagnac interferometer. *Phys. Rev. Lett.*, 81:2446–2449, Sep 1998.

[46] A. V. Sergienko, Y. H. Shih, and M. H. Rubin. Experimental evaluation of a two-photon wavepacket in type-II parametric downconversion. *J. Opt. Soc. Am. B*, 12:859, 1995.

[47] Y. H. Shih and S. A. Sergienko. A two-photon interference experiment using type II optical parametric down conversion. *Phys. Lett. A*, 191:201, 1994.

[48] C.. Silberhorn, P. K. Lam, O. Weiß, F. König, N. Korolkova, and G. Leuchs. Generation of continuous variable Einstein-Podolsky-Rosen entanglement via the Kerr nonlinearity in an optical fiber. *Phys. Rev. Lett.*, 86:4267–4270, May 2001.

[49] A. Sizmann and G. Leuchs. The optical Kerr effect and quantum optics in fibers. *Prog. Opt.*, XXXIX:375, 1999.

[50] H. Takesue and K. Inoue. Generation of polarization-entangled photon pairs and violation of Bell's inequality using spontaneous four-wave mixing in a fiber loop. *Phys. Rev. A*, 70:031802, Sep 2004.

[51] P. Usachev, J. Söderholm, G. Björk, and A. Trifonov. Experimental verification of differences between quantum and classical polarization properties. *Opt. Commun.*, 193:161, 2001.

[52] W. J. Wadsworth, N. Y. Joly, J. C. Knight, T. A. Birks, F. Biancalana, and P. S. J. Russell. Supercontinuum and four-wave mixing with Q-switched pulses in endlessly single-mode photonic crystal fibres. *Opt. Express*, 12:299, 2004.

[53] D. F. Walls and G. J. Milburn. *Quantum optics*. Springer, 2008.

[54] F. Wolfgramm, A. Cerè, F. A. Beduini, A. Predojević, M. Koschorreck, and M. W. Mitchell. Squeezed-light optical magnetometry. *Phys. Rev. Lett.*, 105:053601, Jul 2010.

13 Applications of quantum polarization states

Applications of polarized light in general are so broad that discussing them is outside of this book's scope. We can only mention polarimetry, ellipsometry, interferometry, and devices already discussed in Chapter 9. A detailed review of applications can be found in monograph [12]. Here we only describe two applications of *polarized nonclassical light*, which appeared recently, and therefore should be discussed in connection with the quantum-optical description of polarization. These are tests of the foundations of quantum mechanics and quantum key distribution. While the first subject is purely fundamental, the second one is ultimately applied: quantum key distribution systems are now commercially available. What is common for these subjects is that they both use polarized photons as qubits, i. e., quantum bits of information.

13.1 Bell's inequality and its violation

This section will introduce the reader to just one, among many, implementations of the Bell states. Namely, they play a key role in resolving one of the most disputable situations in quantum mechanics—the Einstein–Podolsky–Rosen (EPR) paradox [15].

13.1.1 The EPR paradox and Bell's inequality

In 1935, Einstein, Podolsky, and Rosen formulated a paradox that seemed to undermine the very basics of quantum mechanics. In the *gedanken* (thought) experiment proposed by EPR, two particles A, B had, at the time of their birth, correlated values of position and anti-correlated values of momentum [15], so that they were born at one point and propagated in the opposite directions. The paradoxical statement, according to the quantum-mechanical idea of measurement, was that upon the measurement performed on particle A, for instance, the measurement of position, the state of particle B would be *instantly* turned into a position state, i. e., into a state with a fixed position. Alternatively, a momentum measurement on particle A would *instantly* put particle B into a momentum state. This *instant state reduction* should happen regardless of the distance between the particles, which could be very large. This meant 'spooky action at a distance', as Einstein formulated it [2], and could not be accepted by most physicists. Moreover, by measuring non-commuting variables, momentum for particle A and position for particle B, one could apparently violate the uncertainty relation.

The argument by EPR went further to propose that the quantum-mechanical description of a quantum system, such as these two particles, was incomplete. The theory, according to their viewpoint, should additionally contain some 'hidden variables', i. e., the *a priori* values of the position and momentum for each particle of the

https://doi.org/10.1515/9783110668025-013

pair. According to the 'Copenhagen' (Bohr's) interpretation, these hidden variables did not exist.

This paradox remained the subject of almost purely philosophic discussions until Bohm [3] proposed to formulate it in terms of binary variables, such as the projection of the spin of a spin-1/2 particle on the direction of the magnetic field. This quantity can be measured in the Stern–Gerlach experiment (Fig. 11.4(a)). One can then imagine two spin-1/2 particles born in such a way that each of them has the spin direction uncertain, but there is perfect anti-correlation between the spin directions of the two particles: if particle A has the spin 'up', particle B has the spin 'down', and vice versa. The state vector of the whole quantum system can be written as

$$|\Psi\rangle = \frac{1}{\sqrt{2}}(|\uparrow\rangle_A |\downarrow\rangle_B + |\downarrow\rangle_A |\uparrow\rangle_B), \tag{13.1}$$

where $|\uparrow\rangle$ denotes a 'spin up' state, and $|\downarrow\rangle$, a 'spin down' state. Note that here the states are distinguished relative to the vertical magnetic field; we could also consider the direction of the spin with respect to the horizontal magnetic field or, actually, any magnetic field direction.

Equation (13.1) gives an example of an entangled state—similar to the $|\Psi^{(+)}\rangle$ Bell state we considered in the previous section: taken separately, each particle has the spin direction completely uncertain, but if particle A has the spin directed 'up', then the spin of particle B is directed 'down', and vice versa. (In fact, the term 'entangled'—in German, 'verschränkt'—has been proposed by Schrödinger namely for this case.) The EPR paradox, in Bohm's formulation, sounds as follows: if one measures the spin direction of particle A, and the result is 'down', then particle B is instantly reduced to the 'spin up' state; but how can this be true if the particles are very far apart? Also, similar to the argument of EPR, one can then simultaneously measure two non-commuting Pauli operators: say, σ_x for particle A and σ_y for particle B.

This binary version of the EPR paradox is extremely useful for two reasons. First, it allows for the derivation of an inequality that can be tested in experiment [2]. (We will derive this *Bell inequality* in the next few paragraphs.) Second, as mentioned in the previous chapters, there is an analogy between the spin of a spin-1/2 particle and the Stokes observables for a photon. The latter can be measured very simply using a setup of Fig. 11.4(b) instead of a bulky complicated setup with magnets, as in the case of the Stern–Gerlach experiment (Fig. 11.4(a)).

Based on the binary interpretation of the EPR paradox, in 1964 John Bell formulated a theorem that assumed the existence of local hidden variables for the two particles and resulted in an inequality. The term 'local' means here that the variables for particle A cannot affect particle B, and vice versa. Because the original derivation of Bell's inequality is a bit complicated [2], here we will present its very simple version in the form of Clauser–Horne–Shimony–Holt (CHSH) inequality, following Klyshko [14].

Let the state of particles A, B be described by binary variables a, b, each of them taking values 1 or –1. Suppose these variables can be measured in an experiment as shown in Fig. 11.4(a) or Fig. 11.4(b): if the upper detector clicks, then we say that $a = 1$, and if the lower detector clicks, then $a = -1$. For the measurement of b we have another similar setup, placed in the path of particle B. Moreover, the same setups in different configurations can measure other variables: a' for particle A and b' for particle B. The variables a' and b' are also binary and take values $+1$ or -1. One can imagine that, for the measurement of a, the setup in the path of particle A looks as shown in Fig. 11.4(a), or in Fig. 11.4(b) with a certain setting of the waveplates. For the measurement of a', the setup has to be modified; for instance, in the case of spin, the setup in Fig. 11.4(a) should be rotated. In the case of the Stokes measurement (Fig. 11.4(b)), the orientations of the HWP and QWP should be changed to measure a'.

To derive Bell's inequality, we assume that, as soon as particles A, B are created, each of them has certain parameters: particle A has parameters a, a' and particle B has parameters b, b'. These are exactly the *local hidden variables* in the EPR argument. For instance, for a spin-1/2 particle these can be projections of the spin on the horizontal and vertical axes. The set of these parameters $\{\lambda\} \equiv \{a, a, b, b'\}$ is assumed to have some probability distribution, $p(\{\lambda\})$. We assume this probability distribution to be 'well-behaved', i. e., to be non-negative and normalized: $p(\{\lambda\}) \geq 0$, $\int p(\{\lambda\})d\{\lambda\} = 1$. In the Copenhagen picture of quantum mechanics, there are no *a priori* values of $\{\lambda\}$, i. e., there are no hidden variables. Let us see where the hidden-variable assumption leads us. Of course there is another assumption here, namely, locality: particle A cannot affect particle B, and vice versa.

We now introduce a new variable,

$$F \equiv \frac{1}{2}\{ab + a'b + ab' - a'b'\} = \frac{1}{2}\{a(b + b') + a'(b - b')\}. \tag{13.2}$$

Because $b, b' = \pm 1$, either $b = b'$ is possible or $b = -b'$; then only one of the round brackets in Eq. (13.2) can be nonzero—and the absolute value of this nonzero bracket is 2. Because $|a| = |a'| = 1$, the new variable F can only take values $+1$ or -1. Therefore F is also binary, and its absolute value is always $|F| = 1$. Let us now look at its mean value. If we try to average F experimentally, every measurement will yield either $F = +1$ or $F = -1$, and the averaging should yield a value $-1 \leq F \leq 1$. An example of such a measurement is shown in Fig. 13.1: the measured points, up to the experimental uncertainty, are either at $F = 1$ or at $F = -1$. By averaging these results over time t or over many experimental tries, we will always get $|\langle F \rangle| \leq 1$.

The same result follows from the calculation using the probability theory,

$$\langle F \rangle = \int F(\{\lambda\})p(\{\lambda\})d\{\lambda\}. \tag{13.3}$$

If the probability distribution $p(\{\lambda\})$ is 'well-behaved', and $|F(\{\lambda\})| \leq 1$, this leads to

$$|\langle F \rangle| \leq 1. \tag{13.4}$$

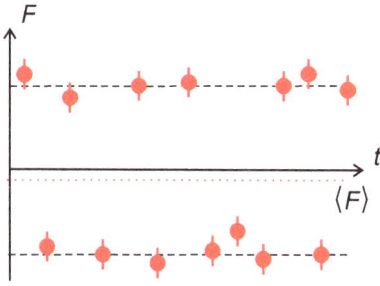

Figure 13.1: Hypothetical measurement of observable F in an experiment. Each experimental result gives values close to either $F = 1$ or $F = -1$. The mean value will always satisfy $|\langle F \rangle| \leq 1$.

In other words, from the assumptions of (i) the existence of hidden variables and (ii) their locality, inequality (13.4) follows. This is a modified [14] Bell inequality in the CHSH form [2, 5, 7]. It is one of the numerous formulations of Bell's inequality, and, like the other Bell inequalities, it can be tested in experiment.

13.1.2 Bell's inequality for Stokes observables

The derivation of Bell's inequality in the previous subsection seems so convincing that it is hard to believe that something is wrong here and that the inequality could be violated. And yet there are situations where it does not hold. In this subsection, we demonstrate such a situation theoretically; in the next subsection we will discuss some experiments. Both the theory of this subsection and the experiments of the next one will deal with polarization-entangled photons.

For calculating the left-hand side of inequality (13.4), we have to choose the binary variables a, a', b, b'. Here we again follow the Klyshko approach, in which the inequality is formulated in terms of the Stokes observables [14]. Consider a Stokes measurement setup like the one shown in Fig. 11.4(b): any Stokes observable for a single photon takes values ± 1, i. e., can be used as one of the binary variables a, a', b, b'. As particles A and B, we will choose the two polarization-entangled photons emitted via SPDC in the directions A and B either under type-II phase matching (Fig. 12.3) or in the interferometric scheme (Fig. 12.4). Let us place a Stokes measurement setup into each path A, B (Fig. 13.2). For testing Bell's inequality, it is sufficient to have only a HWP in each arm, as we will see soon. Let us define the variables a, a', b, b' as follows.

1. For the measurements on photon A, we will choose a to be the first Stokes observable. To measure it (see Chapter 5, Fig. 5.7 or Chapter 11, Fig. 11.4(b)), we need to orient the HWP in path A horizontally. Then, if the 'reflected-path' detector (A1) clicks, we say $a = -1$; if the 'transmitted-path' detector (A2) clicks, we say $a = 1$.

Variable a' will be the second Stokes observable. To measure it, we need to orient the HWP in path A at an angle $22.5°$. Again: if detector A1 clicks, we say that $a' = -1$, and if detector A2 clicks, we say that $a' = 1$. Note that the measurements of a and a'

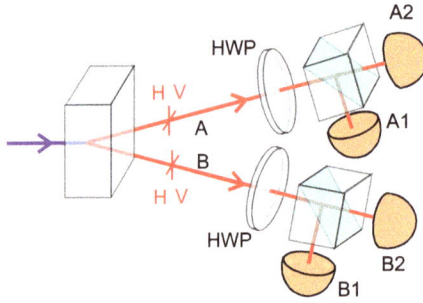

Figure 13.2: Experimental setup for testing Bell's inequalities.

for photon A cannot be performed simultaneously. This is going to be the key point in the explanation why the inequality will be violated.

2. For the measurements on photon B we introduce more complicated variables. Variable b will be the Stokes observable $(S_1 + S_2)/\sqrt{2}$. For this measurement, the HWP in path B should be oriented at $11.25°$ (half-way between the measurement of the first and second Stokes observables). Now, a click of detector B1 will tell us that $b = -1$ and a click of detector B2 will indicate $b = 1$.

For the variable b', we will choose the Stokes observable $(S_1 - S_2)/\sqrt{2}$, and to measure it, we will orient the HWP in path B at an angle $-11.25°$ or, equivalently, at $78.75°$. The clicks of detectors B1 and B2, again, will mean that $b' = -1$ and $b' = 1$, respectively. As in the case of the measurements on photon A, the measurements of b and b' are impossible to perform simultaneously.

Because we are doing a quantum-mechanical calculation, all observables should be operators:

$$\hat{a} \equiv \hat{S}_1^A, \quad \hat{a}' \equiv \hat{S}_2^A, \quad \hat{b} \equiv \frac{1}{\sqrt{2}}(\hat{S}_1^B + \hat{S}_2^B), \quad \hat{b}' \equiv \frac{1}{\sqrt{2}}(\hat{S}_1^B - \hat{S}_2^B), \tag{13.5}$$

where the upper indices A, B mean that the Stokes operators relate to photons A, B.

The variable F will then also be an operator, and take the form

$$\hat{F} = \frac{1}{2}\{\hat{a}(\hat{b} + \hat{b}') + \hat{a}'(\hat{b} - \hat{b}')\} = \frac{1}{\sqrt{2}}\{\hat{S}_1^A \hat{S}_1^B + \hat{S}_2^A \hat{S}_2^B\}. \tag{13.6}$$

Let us now calculate the mean value of $\hat{F}$ by averaging it over the singlet Bell state $|\Psi^{(-)}\rangle$ (12.33). Actually, the inequality will be violated for all four Bell states, but then different operators $\hat{F}$ have to be chosen. With the one chosen according to Eq. (13.6), the inequality will be violated only for $|\Psi^{(-)}\rangle$ and $|\Phi^{(+)}\rangle$.

As the first step, let us show that $|\Psi^{(-)}\rangle$ is an eigenstate of $\hat{F}$. We have to calculate

$$\hat{F}|\Psi^{(-)}\rangle = \frac{1}{2}\{\hat{S}_1^A \hat{S}_1^B + \hat{S}_2^A \hat{S}_2^B\}(|H\rangle_A|V\rangle_B - |V\rangle_A|H\rangle_B). \tag{13.7}$$

We notice that the operators $\hat{S}_{1,2}^A$ act only on photon A, and the operators $\hat{S}_{1,2}^B$, only on photon B. Moreover, the states $|H\rangle$ and $|V\rangle$ are eigenstates of the operator $\hat{S}_1$, with the eigenvalues 1, −1, while $\hat{S}_2$ acts on them as follows: $\hat{S}_2|H\rangle = |V\rangle$, $\hat{S}_2|V\rangle = |H\rangle$. With an account for all this, the algebra becomes very simple, and we obtain

$$\hat{F}|\Psi^{(-)}\rangle = -\sqrt{2}|\Psi^{(-)}\rangle. \tag{13.8}$$

Hence, $|\Psi^{(-)}\rangle$ is indeed an eigenstate of $\hat{F}$, and the mean value $\langle\hat{F}\rangle$ is easy to calculate:

$$\langle\Psi^{(-)}|\hat{F}|\Psi^{(-)}\rangle = -\sqrt{2}, \tag{13.9}$$

and

$$|\langle\hat{F}\rangle| = \sqrt{2} > 1. \tag{13.10}$$

For the state $|\Phi^{(+)}\rangle$, the result is $\langle F\rangle = \sqrt{2}$.

We see now that, contrary to our expectations, the CHSH inequality (13.4) is violated by a quantum-mechanical calculation. Then it should also be violated in experiment, which will be the subject of the next subsection.

13.1.3 Bell's inequality tests

Violation of Bell's inequality, predicted by quantum mechanics, is incompatible with the assumption of local hidden variables, or *local realism*, as one often says. Meanwhile, the term 'realism' is rather vague; today, quantum mechanics is considered to be perfectly realistic, and it correctly predicts the outcome of experiments. In this section, we will very briefly describe the experimental tests of Bell's inequalities.

Instead of variable F, such tests usually operate with the variable [6]

$$S = \frac{1}{2}(\langle\hat{F}\rangle - 1), \tag{13.11}$$

which includes the mean values of the same four terms of Eq. (13.2). For S, the CHSH inequality takes the form

$$-1 \le S \le 0. \tag{13.12}$$

Quantum-mechanical calculation for the state $|\Psi^{(-)}\rangle$ gives $\langle F\rangle = -\sqrt{2}$ and therefore $S = -\frac{1}{2}(\sqrt{2}+1) \approx -1.21$. For the state $|\Phi^{(+)}\rangle$, the result is $S = \frac{1}{2}(\sqrt{2}-1) \approx 0.21$.

In order to test the CHSH inequality (13.12), an experimentalist has to measure the mean values of all terms in Eq. (13.2). As mentioned above, each of them requires a different setting of the setup in Fig. 13.2. For each setting, the correlation of two Stokes observables is measured:

(i) $\frac{1}{\sqrt{2}}\langle\hat{S}_1^A(\hat{S}_1^B + \hat{S}_2^B)\rangle$ is measured with the HWP in channel A at 0° and the HWP in channel B at 11.25°.

(ii) $\frac{1}{\sqrt{2}}\langle \hat{S}_2^A(\hat{S}_1^B + \hat{S}_2^B)\rangle$ is measured with the HWP in channel A at 22.5° and the HWP in channel B at 11.25°.

(iii) $\frac{1}{\sqrt{2}}\langle \hat{S}_1^A(\hat{S}_1^B - \hat{S}_2^B)\rangle$ is measured with the HWP in channel A at 0° and the HWP in channel B at 78.75°.

(iv) $\frac{1}{\sqrt{2}}\langle \hat{S}_2^A(\hat{S}_1^B - \hat{S}_2^B)\rangle$ is measured with the HWP in channel A at 22.5° and the HWP in channel B at 78.75°.

In practice, one only needs to measure the rate of coincidences between the 'transmitted-path' detectors A2 and B2. Indeed, consider the mean value (i). Each of the Stokes observables entering it is measured (see Section 11.2.3) as the difference of the number of photons hitting detector 2 and the number of photons hitting detector 1, for a certain setting of the HWP. Meanwhile, the number of photons N_{A1} hitting detector A1 is equal to $N_A - N_{A2}$, where N_{A2} is the number of photons hitting A2 and N_A is the total number of photons in channel A (which can be measured by removing the polarizing beamsplitter). Then the mean value (i) can be written as

$$
\begin{aligned}
&\frac{1}{\sqrt{2}}\langle \hat{S}_1^A(\hat{S}_1^B + \hat{S}_2^B)\rangle \\
&= \frac{\langle [N_{A2}(0°) - N_{A1}(0°)][N_{B2}(11.25°) - N_{B1}(11.25°)]\rangle}{\langle N_A N_B\rangle} \\
&= \frac{4\langle N_{A2}(0°)N_{B2}(11.25°)\rangle - 2\langle N_A N_{B2}(11.25°)\rangle - 2\langle N_B N_{A2}(0°)\rangle + \langle N_A N_B\rangle}{\langle N_A N_B\rangle},
\end{aligned}
$$

where the angles in brackets denote the orientations of the HWPs. This expression involves only the numbers of transmitted photons and the total photon numbers in channels A, B.

By performing the same calculation for each of the mean values (i)–(iv), we obtain the mean value of $\langle F\rangle$, and then the value of S, in the form

$$
\begin{aligned}
S = \frac{1}{\langle N_A N_B\rangle}[&\langle N_{A2}(0°)N_{B2}(11.25°)\rangle + \langle N_{A2}(22.5°)N_{B2}(11.25°)\rangle \\
&+ \langle N_{A2}(0°)N_{B2}(78.75°)\rangle - \langle N_{A2}(22.5°)N_{B2}(78.75°)\rangle \\
&- \langle N_A N_{B2}(11.25°)\rangle - \langle N_{A2}(0°)N_B\rangle]. \tag{13.13}
\end{aligned}
$$

Experiments on testing the CHSH inequality, according to Eq. (13.13), require the following measurements: four series with polarizing beamsplitters in both arms, two series with polarizing beamsplitters in one arm, and one series with no polarizing beamsplitters. Because the reflected paths are not used, flat polarizers can replace polarizing beamsplitters. Also, real experiments use no HWPs but instead, polarizers oriented at 0°, 22.5°, 45°, 67.5°.

The first photon-based experiments on Bell's inequality violation obtained photon pairs from the cascaded transitions of atoms. Early experiments involved fewer settings of polarizers [9] and rather few statistics, but later, experiments were more and more advanced, and by 1982 the violation of Bell's inequalities was convincingly proved. Especially important were experiments by Aspect et al. [1], where the orientation of the polarizers was varied during the flight of photons from the source to the detectors. This enabled the refutal of several local hidden variables theories—for instance, the hypothesis that after the detection of photon A in a certain polarization state, this information was somehow transmitted to photon B.

Nevertheless, there remained certain 'loopholes' for the local hidden variable theories. One of them, called the *fair sampling loophole*, or detection loophole, is that all, or almost all, pairs should be probed, otherwise there is still space for local hidden variables. Experiments with atomic cascaded transitions still left this loophole open, because atoms emit photon pairs into the full solid angle of 4π radians, and it is very difficult to detect even half of them.

A breakthrough was made when Bell's inequality tests started to use SPDC as a source of photon pairs. In the first experiments of this kind, Shih and Alley [19] and, independently, Ou and Mandel [16], produced entangled states from type-I SPDC, using HOM-type interference on a beamsplitter. Further tests of Bell's inequality were performed with SPDC configurations shown in Fig. 12.3 and Fig. 12.4.

After this pioneering work, Bell tests were repeated always with SPDC, with higher and higher accuracy, and all of them resulted in the violation of Bell's inequalities. And still, some 'loopholes' for local realism remained. First, closing the aforementioned fair sampling loophole required a detection efficiency of at least 82% in the case of maximally entangled states (12.32)–(12.35) and somewhat lower in the case of non-maximally entangled states (with unbalanced terms in the superposition)—but still above 70% for realistic cases. Second, there remained the *locality (communication) loophole*. To exclude the possibility of communication between particles A and B, measurements on A, B should be separated by a spacelike interval in the Minkowski space. This situation is shown in Fig. 13.3: the source S (blue circle) emits photons A, B along the red dashed lines in the Minkowski space. The detection procedures are shown by red rectangles. To avoid communication between the setting of polarizers in paths A, B (green rectangles) and the measurements on the other sides, these events should be also separated by spacelike intervals (green dashed lines in the figure). This requires a relatively fast setting of the polarizers and measurement and a relatively large separation between the measurement stations and the source. Finally,

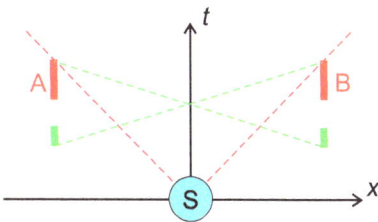

Figure 13.3: Minkowski space diagram for timing in loophole-free tests of Bell's inequalities. The setting of the polarizers in each arm (green rectangle) should be separated from the measurement in the other arm (red rectangle) by a spacelike interval (dashed line).

there existed the *freedom-of-choice loophole*: the settings of the polarizers should be chosen free or random. This implies using some really random choice—a true random-number generator, for instance.

All three loopholes were overcome in recent experiments [11, 13, 18]. The locality loophole was overcome by using large distances between detectors A and B (60 m, 170 m and 1.3 km, respectively) in combination with fast electronics. The freedom-of-choice loophole was eliminated by applying various random-number generators. The fair sampling loophole was closed by using a non-maximally entangled state and detection with 73 % [18] and 76 % [11] efficiencies, provided by superconducting nanowires and transition-edge sensors. Two experiments [11, 18] were performed with entangled photons generated via SPDC and one [13], with spin-1/2 excitations in nitrogen-vacancy centers in diamond, which allowed for a perfect detection efficiency. These three loophole-free tests of Bell's inequality seem to have put an end to the debates about the local hidden variables.

In the end, it is time to explain: why does the Bell inequality (13.12) or (13.4) break down? For many, a good answer will be 'because this is what quantum mechanics says; see the result of calculation (13.10). Photons A, B form a joint quantum system, whose each part is in a mixed state and has zero mean values of all Stokes variable $S_{1,2,3}$ before the measurement.' But the question is then: where did we make a mistake in the derivation of inequality (13.4)? Following again Klyshko's argument [14], the answer is that we assumed the existence of a joint probability distribution $p(\{\lambda\})$ for all variables a, a', b, b' entering the expression for F. Moreover, we assumed this probability distribution to be non-negative. At the same time, we have noticed that different mean values entering inequality (13.4) cannot be measured simultaneously. Therefore, their joint probability distribution is unphysical. We faced a similar situation in Section 11.1.3: the Wigner function pretends to be a joint probability distribution for variables (position and momentum) that cannot be measured simultaneously; the price is that it can be negative. Similarly, we can force Bell's inequality to hold true by allowing the probability distribution $p(\{\lambda\})$ to take negative values [8].

13.2 Quantum key distribution

As already mentioned, a polarized photon represents a superposition state of a two-state quantum system (sometimes one says: two-level quantum system), or a qubit. There are various ways to implement a qubit: for instance, a spin-1/2 particle, already discussed in Chapter 11, a two-level atom, a superconducting circuit. In contrast to these material quantum systems, a photon is much easier to transmit. A photon is therefore an ideal qubit for communication, a carrier of quantum information. But even with photons, there are many ways to implement a qubit: as a path taken in a two-arm interferometer, a phase, or a time interval. *Polarization qubits* are convenient for free-space communication because the polarization state of light is very little affected by propagation in the atmosphere—this is why, as discussed in Chapter 1, light coming directly from the Sun is fully unpolarized, but it gets polarized due to scattering. In this section, we explain how polarized photons can be used—and, actually, are used—to transmit secret messages.

13.2.1 Secret key

People always needed to send secret messages. No matter what the secrets were, military, trade, or personal, it was often important to protect a message against a possible interception. To encrypt a message means to put into correspondence to every letter some symbol, number, or some other letter, to create a cipher. To decrypt the original message, the cipher has to be used again. If the same cipher is used several times, then it can be broken by noticing certain regularities in the encrypted texts. This is exactly how the Enigma machine codes were broken during WW2 [4].

A powerful way to encrypt a message is to use the so-called *one-time pad*, proposed by Vernam. The message should then first be binary encoded, i. e., represented by a sequence of 'zeros' and 'ones'. For instance, the word 'light' is represented by a string of 40 bits (the intervals are added for clarity):

0110 1100 0110 1001 0110 0111 0110 1000 0111 0100

The encryption is done by summing it modulo 2 with the key, i. e., another string of bits, of the same length. For instance, here is a randomly generated code of 40 bits:

0110 0001 0011 0001 1000 0000 0001 1001 0011 0110

The sum of the message and the key modulo 2 is

0000 1101 0101 1000 1110 0111 0111 0001 0100 0010

One who knows the secret key can decrypt the message by summing it with the key modulo 2 again. As proved by Shannon, if the secret key is used only once, has the same length as the message, and is purely random, then the encryption is perfectly secure [4].

Therefore the only task of cryptography is to distribute the secret key between two users, in such a way that it is best protected from an interceptor (an *eavesdropper*). There are many ways to do it—for instance, by simply sending it with some person—but all these ways are vulnerable. The great advantage offered by quantum physics is the 'fragility' of a single quantum system, and the fact that a measurement performed on it should, in general, destroy its state. Therefore in quantum cryptography, the secret key is distributed by imparting the bits to the state of a quantum system, thus turning them into qubits. The qubits can be then physically sent from one user to another. This ground-breaking idea started the whole field of quantum key distribution (QKD), which is now the most industrialized part of quantum information science.

An important part of the whole principle of QKD is that a single quantum system cannot be copied (cloned). This *no-cloning theorem* can be rigorously proved [4] and it means that an eavesdropper cannot copy the transmitted qubits and this way get access to the secret key, or at least to a part of it.

The goal of QKD is to distribute the secret key between two legitimate users, who are traditionally called Alice (A) and Bob (B). In the course of distribution, the key should be protected from the eavesdropper, who is traditionally called Eve (E). The exchange of information involves at least two channels: the *quantum channel*, through which the qubits are sent, and the *public channel*, which could be radio, television, Internet, and which is accessible to everyone. The quantum channel can be noisy; the public channel is more robust: the information transmitted through it cannot be modified.

Below we will only briefly describe the main ways (protocols) of QKD, making an accent on the use of the polarization degree of freedom. Here we provide only the basic principles of each protocol, but for the details we refer the reader to two reviews: one of the earliest, on the basics of the method [10], and the most recent one, on practical QKD [21].

13.2.2 BB84 protocol

The first QKD protocol was proposed in 1984 by Bennett and Brassard [4], and is referred to as BB84, according to the tradition of labeling protocols in cryprography. This was the first implementation of the idea to encode every bit of the secret key into the state of a quantum system. As such, Bennett and Brassard proposed the polarization state of a single photon. As described in Chapter 11, it is impossible to measure two different Stokes observables for a single photon simultaneously. For example, if a photon is diagonally polarized, then in a setup for measuring the first Stokes observable S_1 (Fig. 3.4) it will be reflected or transmitted with 50 % probability. Without knowing how the photon was polarized, an eavesdropper cannot learn it with certainty. Cloning a polarized photon would be also impossible, according to the no-cloning theorem. In an attempt to intercept a qubit, Eve the eavesdropper will have to detect the photon, and then she will either reveal herself, as the photon would not arrive at the receiver, or she will have to re-send this photon. But then she will inevitably send a photon with a wrong polarization state. This will lead to errors, and again Eve will reveal herself.

The protocol then runs as follows. Alice sends a sequence of pulses (for instance, femtosecond pulses with 80 MHz repetition rate), each of which, ideally, contains a single photon polarized differently. Alice encodes each bit of the secret key into the polarization states of these photons. But she uses two different rules for encoding. In half of the cases, she encodes '0's into horizontally polarized photons $|H\rangle$ and '1's into vertically polarized photons $|V\rangle$ (red arrows in Fig. 13.4). But the other half of bits, chosen randomly, are encoded using a diagonal polarization basis (blue arrows in Fig. 13.4). Then, a diagonally polarized photon $|D\rangle$ corresponds to bit '0' and the anti-diagonally polarized photon $|A\rangle$, to bit '1'.

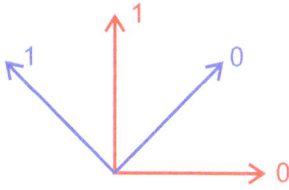

Figure 13.4: Encoding bits of the key in the BB84 protocol. Alice randomly switches between the HV (red) basis or the DA (blue) basis.

In order to encode the bit, Alice could have all photons initially polarized the same way, for instance, horizontally ($|H\rangle$), and then, for each bit, perform a different polarization transformation. In the simplest case, she can rotate a HWP: by $\pm 22.5°$ to prepare photons $|D\rangle$ and $|A\rangle$, and by $45°$ to prepare photons $|V\rangle$. In practice, of course, for increasing the transmission rate the procedure is implemented differently [21].

Let us illustrate the protocol by making a table [4, 10]. The first line in Table 13.1 shows the secret key to be transmitted. As an example, we will use the first 10 bits of the random sequence from the previous section. The second line shows the basis chosen by Alice: 'X' for the diagonal–antidiagonal one and '+' for the horizontal–vertical one, chosen randomly. In the third line, there are states that Alice prepares. They are unambiguously determined by the bit in the first line and the chosen basis in the second line: for instance, bit '0' in the 'X' basis should be $|D\rangle$.

The receiver, Bob, measures the polarization using a standard Stokes measurement setup, as shown in Fig. 11.4 (b) or (c). It is sufficient to have a single HWP, selecting either the measurement of S_1 ('horizontal–vertical' measurement basis), or the measurement of S_2 (the 'diagonal–antidiagonal' basis). As in the case of transmission, real-life setups do not use rotation of waveplates but faster methods to switch between the measurements in different polarization bases. Bob does not know, in which basis a bit was encoded; therefore he randomly chooses the basis (the fourth line in Table 13.1). This way, Bob unambiguously distinguishes between H and V polarizations if he uses the '+' basis. But in approximately half of the cases, he uses the 'X' basis

Table 13.1: Example of a BB84 protocol [4].

Random bit	0	1	1	0	0	0	0	1	0	0
Alice's basis	+	+	X	+	+	X	X	+	X	+
Alice's qubit	$|H\rangle$	$|V\rangle$	$|A\rangle$	$|H\rangle$	$|H\rangle$	$|D\rangle$	$|D\rangle$	$|V\rangle$	$|D\rangle$	$|H\rangle$
Bob's basis	+	X	X	X	+	X	X	+	X	X
Bob measures	H	D	A	A	H	D	D	V	D	A
Same basis?	Y	N	Y	N	Y	Y	Y	Y	Y	N
Sifted key	0		1		0	0	0	1	0	
Test Eve?					Y				Y	
Secret key	0		1			0	0	1		

and then, makes a mistake if the qubit was $|H\rangle$ or $|V\rangle$. The conclusions Bob makes are shown in the fifth line of the table.

After a certain number of bits have been transmitted (and all photons have been detected and destroyed!), Bob publicly announces which basis he used for each bit. Alice then says in which cases they used the same bases. Alice and Bob discard the bits where they used different bases, and leave only those where they used the same ones. After this procedure, called the *key sifting*, the length of the key is reduced approximately twice, because the probability for Alice and Bob to use the same basis is 50 %. Nevertheless, the part of the key that remains is random, and long enough if sufficiently many bits have been sent. The key is common for Alice and Bob, as long as there were no errors during the transmission. These could be caused both by depolarization or loss of the photons and by the presence of Eve. Indeed, Eve could have 'stolen' some qubits, reproduced them somehow (with errors of course) and then, after the public announcement of the bases used, acquired some part of the random key.

Therefore, Alice and Bob need to check whether Eve interfered in the course of transmission. To this end, they take a part of the key, for instance, (10 %) and compare it (line 8 of the table). This procedure is also made through the public channel, but these 10 % of the key are then discarded. If the eavesdropping took place, the key would contain more errors than losses and depolarization would cause. Then, the whole key is thrown out and the procedure is repeated anew. Otherwise (as shown in the table), the rest of the key is kept and, after applying error correction procedures [4, 10], used as the secret key.

The BB84 protocol can be further advanced by using, apart from the four linear polarized states $|H\rangle$, $|V\rangle$, $|D\rangle$, $|A\rangle$, also circularly polarized photons $|R\rangle$, $|L\rangle$. Various other modifications, as well as the strategies of Eve, are described in Ref. [10]. Current state of the art in QKD can be found in Ref. [21].

An important question is how to produce single photons for QKD. Ideally, the transmitted states should not contain multiphoton components; otherwise the eavesdropper can tap off and use one of the photons. Up to recently, most QKD systems operated with weak coherent pulses, with the mean photon number per pulse below 0.2. Then, the probability to have two photons per pulse is small. However, this probability is still nonzero; in addition, there are 'empty' pulses, which strongly reduces the rate of transmission. One way to obtain single photons is to use single-photon emitters such as atoms, molecules, color centers in diamond, or quantum dots; another way is to obtain single photons from SPDC or FWM through heralding (Section 12.3.2). State-of-the-art QKD methods use both these techniques [21].

13.2.3 EPR-based protocols

Unlike in BB84 protocol, where the qubits are prepared by the sender (Alice) and detected by the receiver (Bob), in EPR-based protocols a single distributor sends the qubits to the two users [10]. The qubits should form an entangled state—one of the four Bell states (12.32)–(12.35); see Section 12.3.3. Both Alice and Bob have Stokes measurement setups and do the same as Bob in the BB84 protocol: they randomly choose bases and write down a zero or unity, depending on whether their photon gets reflected or transmitted.

Two-qubit BB84 protocol. In the simplest case, the state is $|\Phi^{(+)}\rangle$, and Alice and Bob randomly and independently switch between 'horizontal-vertical' and 'diagonal–antidiagonal' bases. If they both use the '+' basis, they have perfect correlation. Either both photons are horizontally polarized and get reflected in both setups—then both Alice and Bob record a '0' bit—or both photons are vertically polarized and get transmitted; then both Alice and Bob write down '1'. But whenever they switch to the 'X' basis, the situation remains the same, because the $|\Phi^{(+)}\rangle$ state is invariant to linear polarization rotation; in the AD basis it takes the same form:

$$|\Phi^{(+)}\rangle = \frac{1}{\sqrt{2}}(|H\rangle_A|H\rangle_B + |V\rangle_A|V\rangle_B) = \frac{1}{\sqrt{2}}(|D\rangle_A|D\rangle_B + |A\rangle_A|A\rangle_B). \qquad (13.14)$$

This symmetry can be seen, for instance, for the macroscopic analogue of the $|\Phi^{(+)}\rangle$ state in Fig. 12.13. Then, if Alice and Bob use the same bases, they have exactly the same bits of their key. The rest of the protocol works similar to BB84.

The E91 protocol, proposed by Ekert in 1991, uses the singlet Bell state $|\Psi^{(-)}\rangle$. As discussed in Section 12.3.3, this state is invariant to any polarization transformation. Therefore, whatever polarization bases Alice and Bob use, they will get the same bits as long as the bases are the same.

In the E91 protocol, they randomly switch between three polarization bases: the 'HV' one, the 'AD' one, and the third one, in which the polarization states are also linear but the polarization directions are at 22.5° and 112.5°. In the Stokes measurement setup of Fig. 11.4 (b), this basis is accessed with no QWP and the HWP oriented at 11.25°; this setting corresponds to the measurement of observable $\frac{1}{\sqrt{2}}(S_1 + S_2)$.

As in the other protocols, after receiving some number of bits, Alice and Bob publicly discuss the bases they used and sift the key by discarding all cases where they used different bases. But in contrast to other protocols, now they can check the existence of an eavesdropper by testing Bell's inequalities with the data they have.

EPR-based protocols were further developed into so-called device-independent protocols; for more details; see Ref. [21].

13.2.4 B92 protocol

The B92 protocol, proposed by Bennett in 1992, uses two non-orthogonal states for QKD. For instance, Alice uses $|H\rangle$ to encode '0' and $|D\rangle$ to encode '1' (Fig. 13.5).

As in the other protocols, Bob chooses his basis randomly, between '+' and 'X'. If he has the photon reflected in the '+' basis, he makes the conclusion that the bit

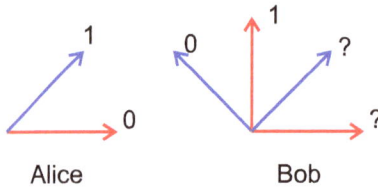

Figure 13.5: Encoding (Alice) and decoding (Bob) bits of the key in the B92 protocol.

was '1': if it were '0', then the photon would be horizontally polarized and would be definitely transmitted. But if with the same basis used, Bob has the photon transmitted, no conclusion can be made. The photon could be either diagonally polarized or horizontally polarized in this case. Therefore, Bob says that the result is inconclusive (Fig. 13.5) and discards this bit.

The same happens if Bob uses the 'X' basis and has a photon transmitted: it could be diagonally polarized, but it also could be horizontally polarized; therefore the result is inconclusive (Fig. 13.5) and the bit is discarded. And only if Bob gets the photon reflected in the 'X' basis, he writes down '0' because the photon could not be diagonally polarized.

The B92 protocol is easier to realize than BB84. Besides, as we will now see, it can be applied to continuous-variable states.

Any two non-orthogonal states can be used for the B92 protocol. It is important though to have states orthogonal to them, or at least approximately orthogonal. An example is a set of two coherent states $|\alpha\rangle$ and $|\beta\rangle$. Two coherent states are always non-orthogonal: their scalar product is $\langle\alpha|\beta\rangle = \exp\{-|\alpha - \beta|^2\}$. For two weak coherent states, this scalar product differs considerably from zero; therefore two weak coherent states are always non-orthogonal.

In the simplest continuous-variable QKD (CV QKD) protocol with discrete modulation [21], Alice encodes the qubits into two coherent states that differ only by the phase: the '0' is encoded into state $|-\alpha\rangle$ and the '1', into state $|\alpha\rangle$ (Fig. 13.6). This encoding is very simple in practice and requires only a phase modulator. The amplitude of the coherent state should be $|\alpha| < 1$.

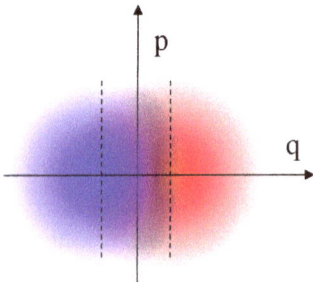

Figure 13.6: Continuous-variable B92 protocol: the '0' is encoded into coherent state $|-\alpha\rangle$ (blue) and the '1', into coherent state $|\alpha\rangle$ (red).

To measure these states, Bob uses homodyne detection. If the value of the q quadrature exceeds the one shown by the right-hand dashed line in Fig. 13.6, the conclusion is that the state was $|\alpha\rangle$, and Bob writes down '1'; if the value is lower than shown by the left-hand dashed line, then the state must have been $|-\alpha\rangle$, and Bob writes down '0'. In all other cases the result is inconclusive. The protocol can be improved by using quadrature-squeezed, rather than coherent states [20].

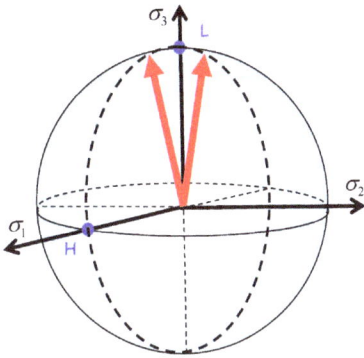

Figure 13.7: Continuous-variable QKD protocol with bright polarized beams. The two states encoding '0' and '1' are shown on the Poincaré sphere with red arrows.

In connection with the main subject of this book, let us mention in the end that the same CV QKD protocol can be realized with polarized bright states of light. Indeed, consider now two strong coherent beams polarized approximately circularly, but with small deviations towards the H and V directions (red arrows in Fig. 13.7). Such states can be shown on the Poincaré sphere with a radius given by the mean photon number. Both states will be close to the North pole, and provided the radius of the sphere is large, the landscape around the states will be practically flat. This situation has been described in Section 11.3; see Fig. 11.7. In the neighborhood of the North Pole, the two bright polarized states will look similarly to the two coherent states of Fig. 13.6: they will be displaced in the H-V direction and partially overlap. A Stokes measurement will distinguish them partially, like the two weak coherent states. In contrast to homodyne detection, here one does not need a local oscillator, since the role of a local oscillator is played by the circularly polarized polarization component. Recently, this protocol was implemented with polarization squeezed states [17].

Bibliography

[1]   A. Aspect, J. Dalibard, and G. Roger. Experimental test of Bell's inequalities using time-varying analyzers. *Phys. Rev. Lett.*, 49:1804–1807, Dec 1982.

[2]   J. S. Bell. *Speakable and unspeakable in quantum mechanics*. Cambridge University Press, 1987.

[3]   D. Bohm. *Quantum theory*. Prentice-Hall, 1952.

[4] D. Bouwmeester, A. Ekert, and A. Zeilinger. *The physics of quantum information*. Springer-Verlag, 2000.

[5] J. F. Clauser and M. A. Horne. Experimental consequences of objective local theories. *Phys. Rev. D*, 10:526–535, Jul 1974.

[6] J. F. Clauser, M. A. Horne, A. Shimony, and R. A. Holt. Proposed experiment to test local hidden-variable theories. *Phys. Rev. Lett.*, 23:880–884, Oct 1969.

[7] J. F. Clauser and A. Shimony. Bell's theorem. Experimental tests and implications. *Rep. Prog. Phys.*, 41(12):1881–1927, dec 1978.

[8] N. V. Evdokimov, D. N. Klyshko, V. P. Komolov, and V. A. Yarochkin. Bell's inequalities and EPR–Bohm correlations: working classical radiofrequency model. *Phys. Usp.*, 39(1):83–98, jan 1996.

[9] S. J. Freedman and J. F. Clauser. Experimental test of local hidden-variable theories. *Phys. Rev. Lett.*, 28:938–941, Apr 1972.

[10] N. Gisin, G. Ribordy, W. Tittel, and H. Zbinden. Quantum cryptography. *Rev. Mod. Phys.*, 74:145–195, Mar 2002.

[11] M. Giustina, M. A. M. Versteegh, S. Wengerowsky, J. Handsteiner, A. Hochrainer, K. Phelan, F. Steinlechner, J. Kofler, J.-A. Larsson, C. Abellán, W. Amaya, V. Pruneri, M. W. Mitchell, J. Beyer, T. Gerrits, A. E. Lita, L. K. Shalm, S. W. Nam, T. Scheidl, R. Ursin, B. Wittmann, and A. Zeilinger. Significant-loophole-free test of Bell's theorem with entangled photons. *Phys. Rev. Lett.*, 115:250401, Dec 2015.

[12] D. Goldstein. *Polarized light*. GRC, 2003.

[13] B. Hensen, H. Bernien, A. E. Dreáu, A. Reiserer, N. Kalb, M. S. Blok, J. Ruitenberg, R. F. L. Vermeulen, R. N. Schouten, C. Abellań, W. Amaya, V. V. Pruneri, M. W. Mitchell, M. Markham, D. J. Twitchen, D. Elkouss, S. Wehner, T. H. Taminiau, and R. Hanson. Loophole-free Bell inequality violation using electron spins separated by 1.3 kilometres. *Nature*, 526:682, 2015.

[14] D. N. Klyshko. Basic quantum mechanical concepts from the operational viewpoint. *Phys. Usp.*, 41(9):885–922, 1998.

[15] L. Mandel and E. Wolf. *Optical coherence and quantum optics*. Cambridge University Press, 1995.

[16] Z. Y. Ou and L. Mandel. Violation of Bell's inequality and classical probability in a two-photon correlation experiment. *Phys. Rev. Lett.*, 61:50–53, Jul 1988.

[17] C. Peuntinger, B. Heim, C. R. Müller, C. Gabriel, C. Marquardt, and G. Leuchs. Distribution of squeezed states through an atmospheric channel. *Phys. Rev. Lett.*, 113:060502, Aug 2014.

[18] L. K. Shalm, E. Meyer-Scott, B. G. Christensen, P. Bierhorst, M. A. Wayne, M. J. Stevens, T. Gerrits, S. Glancy, D. R. Hamel, M. S. Allman, K. J. Coakley, S. D. Dyer, C. Hodge, A. E. Lita, V. B. Verma, C. Lambrocco, E. Tortorici, A. L. Migdall, Y. Zhang, D. R. Kumor, W. H. Farr, F. Marsili, M. D. Shaw, J. A. Stern, C. Abellán, W. Amaya, V. Pruneri, T. Jennewein, M. W. Mitchell, P. G. Kwiat, J. C. Bienfang, R. P. Mirin, E. Knill, and S. W. Nam. Strong loophole-free test of local realism. *Phys. Rev. Lett.*, 115:250402, Dec 2015.

[19] Y. H. Shih and C. O. Alley. New type of Einstein–Podolsky–Rosen–Bohm experiment using pairs of light quanta produced by optical parametric down conversion. *Phys. Rev. Lett.*, 61:2921–2924, Dec 1988.

[20] C. Weedbrook, S. Pirandola, R. García-Patrón, T. C. Ralph, J. H. Shapiro, and S. Lloyd. Gaussian quantum information. *Rev. Mod. Phys.*, 842:621–669, May 2012.

[21] F. Xu, X. Ma, Q. Zhang, H.-K. Lo, and J.-W. Pan. Secure quantum key distribution with realistic devices. *Rev. Mod. Phys.*, 92:025002, May 2020.

Index

analytic signal 7, 8, 12, 14, 36
angle of anisotropy 26, 27, 39, 40
angular spectrum 82–84, 91
anisotropy 24, 25, 38, 39, 105
anti-bunching 143, 145

balanced homodyne detection 144, 145, 215
beam displacer 105, 106
Bell states 182, 183, 190–192, 195, 201, 204, 213
Bell's inequality 200–203, 206–208
Bell's inequality tests 205, 207, 208
Berry phase 53, 62, 63
biaxial crystals 32, 34–36, 39, 123
biphoton 177–179, 181, 182
birefringence 24, 27, 29, 30, 32, 36, 37, 46, 76, 101, 103, 108, 111, 112, 127–129, 172
Brewster's law 107–109, 111
bunching parameter 142, 143, 145, 178, 183

calcite 2, 3, 22, 39, 40, 105, 108, 110, 112, 129
characteristic function 143, 157
circular birefringence 36, 37, 46, 104
coherence matrix 18, 196
coherent state 139–141, 144, 158–160, 174, 187–189, 214, 215
commutation relations 147, 185
commutation relations for the Stokes operators 147
crystal symmetry 31, 32, 121
crystal symmetry classes 31, 32, 121–123, 125, 131

degree of polarization 17, 20, 22, 52, 147, 148, 179, 193, 196, 197
dielectric permittivity 24, 32, 37, 117, 121
dielectric tensor 24, 25, 34, 36, 38, 40
dipole 86, 87, 91
double refraction 2, 3, 29

eavesdropper 5, 209, 210, 212, 213
eigenstate 137–140, 149–151, 154, 155, 204, 205
eigenvalue 138–140, 150, 151, 153–155, 205
electric displacement field 8, 24, 25, 29, 30, 116
entanglement 146, 180, 183, 189, 192
EPR paradox 5, 200, 201
evanescent 81, 86–88, 91, 96

Faraday rotator 105
Fock state 138–140, 143, 149, 150, 177–182
Fresnel equation 29, 32, 34, 35, 38
Fresnel surface 30–38

geometric phase 53, 56, 62, 63
Glan prism 107–111
Glauber–Sudarshan quasi-probability 143, 144
Glauber's correlation functions 141–143, 145, 178, 183
group velocity 27, 29, 31, 39, 40

half-wave plate 44–46, 48–52, 58, 61, 62, 75, 76, 101–103, 106, 111, 112, 152, 153, 158, 170, 175, 180–183, 193–195, 202–206, 211, 213
Hamiltonian 137, 163, 165–167, 169–171, 175–177, 182, 184–188, 190
Hanbury Brown–Twiss experiment 141–144, 178, 183
Heisenberg picture 183, 184, 189
Hermite–Gaussian 67–69, 72, 80, 84
hidden polarization 168, 179, 193–196
high-gain parametric down-conversion 185
Hong–Ou–Mandel effect 180, 181, 188, 193

isotropic crystals 32, 117, 121, 124, 125

Jones matrices 4, 15, 16, 20, 21, 43–47, 181
Jones vector 11–16, 18–20, 42–46, 56–58, 61, 63, 92, 149, 165

Kerr squeezing 174–176, 192, 196
Kleinman's symmetry 120, 121, 123

Laguerre–Gaussian 67–70, 72, 73, 76, 97
linear susceptibility 9, 24, 116, 117
liquid crystal 38, 75, 76, 113

macroscopic Bell states 191, 192
Maxwell's equations 8, 26
Möbius strip 97, 98
modulation instability 171, 173, 175
Mueller matrices 20, 21, 52

Near-field scanning optical microscopes (NSOM) 90
Neumann's principle 31, 121
non-critical phase matching 133

nonclassical state 141, 142, 144, 145, 156, 159, 168, 171, 173, 178, 183, 186, 187, 189, 193, 195
nonlinear susceptibility 116, 117, 119, 120, 126, 130, 164, 165, 170
normal ordering 143

optical activity 36–38, 45, 105
optical isolator 105
orthogonality of polarization states 3, 4, 13, 14, 17, 19, 30, 56–58, 112, 132, 135–138, 142, 150, 154–156

P-distribution 143–145, 159
Pancharatnam phase 56, 59–63
parametric down-conversion 163–172, 175, 177–179, 182, 183, 185, 186, 191, 194, 195, 203, 207, 208, 212
parametric gain 176–178, 182–186, 189–191
periodically poled crystal 131
periscope 55
phase plate 42, 44, 46–48, 50, 52, 63, 183
phase space 143, 156, 158, 174, 184, 185
phase velocity 26, 27, 29, 30, 39, 42
photon annihilation operator 137, 139, 160, 181, 185
photon creation operator 137, 138, 165, 167, 170, 177, 181
photon-number operators 135, 138
Poincaré sphere 4, 16, 18–21, 47–52, 56, 57, 60–63, 72, 73, 114, 157, 159, 160, 179, 197
polarization basis 14, 15, 43, 46, 51, 136, 137, 139
polarization ellipse 10, 11, 13, 14, 19, 20, 73, 92, 96, 97
polarization entanglement 192
polarization modes 5, 16, 17, 22, 111, 135–138, 142, 146, 149, 160, 167, 168, 175, 177, 178, 180, 181, 184–187, 189, 190
polarization of matter 24, 116, 118, 123, 124, 126, 127, 130, 131, 163
polarization prism 22, 51, 58, 60, 107, 112
polarization quasi-probability 156–159, 187, 189–191
polarization squeezing 146, 159, 160, 168, 183, 186–189, 192
polarization-maintaining fiber 111, 112, 176
polarization-scalar light 191, 195
positive-frequency field 7, 26, 136–138, 141

Poynting vector 26, 27, 30, 31, 39, 40, 105
projective measurement 149, 152, 153

q-plate 76, 113–115
quadrature operators 138, 140, 143, 160, 184
quantum key distribution (QKD) 1, 2, 5, 153, 179, 200, 209, 210, 212, 214, 215
quantum measurement of the Stokes observables 150–155, 183, 192, 194, 203, 204, 212, 213
quarter-wave plate 21, 22, 41, 42, 45, 49–52, 58, 102, 103, 106, 112, 152, 158, 183, 193, 202, 213
quasi-phasematching 131
quasi-probabilities 143, 156, 157
qubit 1, 5, 149, 200, 208–214

rotator 45, 46, 50, 51, 60–62, 104, 105

Sagnac interferometer 176, 189
second-order correlation function 141, 142
secret key 5, 209–212
shot noise 140, 159, 174, 187, 189, 190, 195
shot-noise limit 140, 174, 187, 188, 190, 191
SLM 76
spatial light modulator 38, 101, 112, 113
spin 86, 93
spontaneous four-wave mixing 171, 172, 175
SPP 96
squeezed state 140, 141, 174, 188, 215
squeezed vacuum 141, 184, 185, 187–189, 196
Stern–Gerlach experiment 151, 201
Stokes measurement 21, 23, 51, 88, 106, 145, 158, 192, 193, 202, 203, 211, 213, 215
Stokes observable 16, 17, 21, 40, 41, 51, 135, 146, 148, 149, 151–153, 155–160, 183, 186–196, 201, 203–205, 208, 210
Stokes operators 135, 147, 148, 150, 151, 153, 156, 160, 186, 204
Stokes parameters 91–93
Stokes space 157–159, 187, 189, 191
Stokes vector 17–22, 42, 47–49, 51, 63, 91
structured light 65, 67, 70, 73, 75, 86, 87

total internal reflection 81, 87
transverse spin 93–97
twin-beam squeezing 146, 187
type-I phase matching 127–129, 132, 133, 167–170, 178, 188, 191, 207

type-II phase matching 129, 132, 167–170, 177–179, 182, 185–187, 194, 203

uncertainty 140, 146, 153, 159, 166, 174, 185, 202
uncertainty relations 147, 148, 189, 192, 200
uniaxial crystals 32, 33, 35, 36, 38–40, 128, 129

vacuum state 140, 164, 177, 184, 189, 196

walk-off 39–42, 105, 130, 132, 133, 153, 155
walk-off compensation 132
waveplate 42, 50, 51, 75, 76, 101, 104
weak measurement 152, 153, 155, 156
weak value 154–156
Wigner function 141, 144–146, 158, 159, 174, 208